Advanced Power Electronics Technology

Advanced Power Electronics Technology

Guest Editors

Wenzhong Ma
Xingtian Feng
Shuguang Song

MDPI

Basel • Beijing • Wuhan • Barcelona • Belgrade • Novi Sad • Cluj • Manchester

Guest Editors
Wenzhong Ma
School of Shida Shanneng
Renewable Energy
China University of
Petroleum (East China)
Qingdao
China

Xingtian Feng
School of Shida Shanneng
Renewable Energy
China University of
Petroleum (East China)
Qingdao
China

Shuguang Song
School of Shida Shanneng
Renewable Energy
China University of
Petroleum (East China)
Qingdao
China

Editorial Office
MDPI AG
Grosspeteranlage 5
4052 Basel, Switzerland

This is a reprint of the Special Issue, published open access by the journal *Energies* (ISSN 1996-1073), freely accessible at: https://www.mdpi.com/journal/energies/special_issues/10IT914P94.

For citation purposes, cite each article independently as indicated on the article page online and as indicated below:

Lastname, A.A.; Lastname, B.B. Article Title. *Journal Name* **Year**, *Volume Number*, Page Range.

ISBN 978-3-7258-4789-1 (Hbk)
ISBN 978-3-7258-4790-7 (PDF)
https://doi.org/10.3390/books978-3-7258-4790-7

Contents

About the Editors

Wenzhong Ma

Wenzhong Ma received the B.S. and M.S. degrees from the Harbin Institute of Technology, Harbin, China, in 1993 and 1995, respectively, and the Ph.D. degree from the Institute of Electrical Engineering, Chinese Academy of Sciences, Beijing, China, in 2006, all in electrical engineering. He was a visiting professor at the University of Alberta, Edmonton, AB, Canada, from 2013 to 2014. Since 1995, he has been a faculty member at the China University of Petroleum, Qingdao, China, where he is currently a professor in the Department of Electrical Engineering. He has conducted extensive research in electrical engineering and automation, including power electronic systems, converters, renewable energy, distributed microgrid, HVdc, motor design, and motor drives. From 2002 to 2006, he was involved in a National Key Project for Shanghai High-Speed Maglev Train Systems, which is the first commercial high-speed maglev train. He took charge of the commissioning and testing work of the long stator linear motor, propulsion system, and power distribution systems. He holds six patents in his areas of interest. He has authored six books and authored or coauthored more than 30 technical papers. Dr. Ma is a Committee Member of the China Electrotechnical Society.

Xingtian Feng

Xingtian Feng received B.S. and M.S. degrees in electrical engineering from the China University of Petroleum (East China), Qingdao, China, in 2001 and 2004, respectively, and the Ph.D. degree in electrical engineering from the Institute of Electrical Engineering, Chinese Academy of Sciences, Beijing, China, in 2012. Since 2004, he has been with the China University of Petroleum (East China), Qingdao, China. His research interests include power electronics, power quality, distributed generation, and energy storage technology.

Shuguang Song

Shuguang Song received his B.S. degree and Ph.D. degree from Xi'an Jiaotong University, Xi'an, China, in 2014 and 2021, respectively, both in electrical engineering. He is currently a lecturer with the College of New Energy, China University of Petroleum (East China). His current research interests include modeling and control of multilevel converters, HVDC, and power quality.

energies

MDPI

Article

Adaptive Quasi-Super-Twisting Sliding Mode Control for Flexible Multistate Switch

Wenzhong Ma [1], Xiao Wang [1,*], Yusheng Wang [2], Wenyan Zhang [1], Hengshuo Li [1] and Yaheng Zhu [1]

[1] College of New Energy, China University of Petroleum (East China), Qingdao 266580, China; mawenzhong@126.com (W.M.); 19960009@upc.edu.cn (W.Z.); 19930065@upc.edu.cn (H.L.); mawenzhg@163.com (Y.Z.)

[2] PetroChina Planning and Engineering Institute, Beijing 100083, China; wenning99@tom.com

* Correspondence: wxiao200068@163.com

Abstract: The mathematical model of a flexible multistate switch (FMSS) exhibits nonlinear and strong coupling characteristics, whereas traditional power decoupling control makes it difficult to completely decouple the output power. The traditional proportional–integral control parameters are difficult to adjust, and their robustness and dynamic performance are poor, which affects the stability of the voltage of the power distribution network and feeder power. To address these problems, this study first converted the original system into a linear system via coordinate transformation using feedback-accurate linearization to decouple active and reactive currents. Thereafter, a super-twisting sliding mode control (ST-SMC) algorithm was introduced, and an adaptive quasi-super-twisting sliding mode control (AQST-SMC) algorithm comprising the quasi-sliding mode function and adaptive proportional term was proposed. An FMSS double closed-loop controller was designed to achieve improved vibration suppression and convergence speed. A three-port FMSS simulation model was developed using MATLAB/Simulink, and the simulation results show that the proposed control strategy enhances the robustness and dynamic performance of the system.

Keywords: flexible distribution network; flexible multistate switch; exact feedback linearization; adaptive quasi-super-twisting sliding mode control

1. Introduction

With the generalization of distributed energy sources and new types of load access to power systems, traditional distribution networks are experiencing problems such as voltage overruns, bidirectional tidal currents, and feeder power imbalances [1–3]. Flexible multistate switches (FMSSs), widely used in medium-voltage distribution grids, can realize continuous and precise power control, improve feeder voltage distribution, and increase the reliability of grid power supply. Furthermore, they can realize a flexible closed-loop operation of the distribution system and increase the ability to consume distributed energy resources [4,5]. Recently, many related studies on FMSSs have been conducted. The proportional–integral (PI) control strategy has been applied [6]; however, this presents challenges, such as difficult parameter adjustment and weak anti-interference ability of multiport FMSSs. In a previous study [7,8], instead of PI control, finite-set model predictive control (MPC) was proposed, which showed excellent dynamic characteristics and anti-interference in a steady state along with a high current response speed. However, such a model presents challenges, such as unstable switching frequency, high control system computation, and model mismatch. Another study [9] proposed an FMSS-based fault detection method for distribution networks under grid-side AC faults, in which the traditional fault management devices were replaced, with the method achieving better dynamic performance during faults. However, the double closed loop still uses PI control, which is parameter-sensitive and insufficiently robust.

FMSSs belong to nonlinear systems exhibiting a strong coupling degree, and it is challenging to design FMSS controllers by applying traditional control theory. When feedback linearization is employed, a nonlinear system can be linearized by selecting an appropriate coordinate mapping method, which can completely decouple the state variables of the system, thus considerably improving the control performance of the system [10]. Sliding mode control (SMC) is a nonlinear control method that requires a model exhibiting high accuracy but considerably short response time and high robustness; it has been widely used in the design of FMSS controllers and observers. In a previous study, second-order SMC was combined with a nonlinear disturbance observer to design a strategy for suppressing direct current (DC)-side voltage fluctuations caused by the voltage imbalance of FMSS ports [11], which resulted in accurate tracking of the reference voltage and current. However, the nonlinear disturbance observer was sensitive to the system model error and parameter perturbation, and exhibited poor robustness. A soft open point (SOP) strategy based on adaptive voltage drop outer-loop control and feedback-linearized sliding mode inner-loop control was proposed to fully harness the regulation capability of the converter in the SOP system to avoid reaching the voltage limit [12]. Furthermore, the strategy was proposed to balance the tidal current in the distribution network. However, the inner loop uses a conventional first-order linear sliding mode, which may not be sufficiently robust to external perturbations and uncertainties, leading to a decline in the control system's performance. A full-order terminal SMC method was proposed to enhance system immunity to disturbances by establishing a mathematical FMSS model considering parameter uptake [13]. However, the full-order terminal sliding mode response was slower in the initial moments and cases of large nonlinear disturbances. The super-twisting SMC (ST-SMC) algorithm is a second-order sliding mode algorithm. This algorithm only uses information about the system state and its first-order derivatives in designing the system control law, and can be directly used to design controllers and observers for relative-degree-one systems, which makes it widely applicable in second-order sliding mode control [14]. Moreover, it is more accurate and robust than first-order SMC. Furthermore, extensive research has been conducted on the mathematical computation of ST-SMC algorithms. Xu et al. proposed a variable-step closed-loop angle compensation method using the tangent reaching law based on SMC and the convergence law. Results revealed that the algorithm exhibited good dynamic compensation performance and disturbance resistance while maintaining good steady-state compensation accuracy [15]. Maged et al. proposed a new application of ST-SMC based on an artificial hummingbird optimization technique, in order to enhance the transient performance of islanded microgrids [16]. Ghazi et al. proposed an ST algorithm based on the circle search algorithm to improve the efficiency of grid-connected photovoltaic systems; its parameters were optimized in combination with the gray wolf optimizer to improve the robustness of the system, tracking speed, and the optimal convergence to the minimum error value [17].

In recent years, ST-SMC has been widely used in power electronics and power drives; moreover, considerable research has focused on improving the traditional ST-SMC by combining it with other nonlinear control algorithms to optimize the dynamic and steady-state characteristics of the system. Saadaoui et al. proposed an ST-SMC method for ultrafast chargers of pure electric vehicles for grid imbalance conditions, which can always achieve the lowest performance index compared with other nonlinear methods, demonstrating the fast response speed, strong robustness, and low steady-state error of ST-SMC [18]. Celik et al. designed a Kalman filter using an ultra-twisted sliding-mode strategy, which ensures a shallow DC-link voltage ripple and achieves a fast dynamic response for a battery electric vehicle charger based on a three-phase two-level rectifier and a conventional DC/DC buck-boost converter [19]. Tiwary et al. designed an ST-SMC controller based on an isolated DC–DC bidirectional dual active bridge (DAB) power converter, in order to achieve DC bus voltage stability and high robustness for accurate reference voltage tracking [20].

To enhance the performance of the control system in complex environments, some improved ST-SMCs have been created. Wang et al. proposed a model-free predictive current control strategy based on an adaptive super-twisting sliding-mode observer, in order to design an adaptive gain matrix and dynamically adjust the observer parameters to suppress the effect of boundary uncertainty in the dynamic part of the rectifier system, and to ensure the accuracy of the prediction model [21]. Pati et al. proposed adaptive ST-SMC and dual-loop control for a three-phase differential boost inverter in a grid-connected photovoltaic system to regulate voltage across a DC bus capacitor, and to compensate for harmonics under nonlinear loads [22]. However, for FMSS systems, ST-SMC exhibiting improved efficiency should be developed to withstand various complex operating conditions, and exhibit enhanced dynamic performance in complex environments. Composite second-order SMC was proposed to apply ST-SMC in the outer loop of an FMSS [23]. Nevertheless, the full-order fast terminal SMC used in the inner loop requires high model accuracy, is sensitive to measurement errors, and is prone to cause instability in the controller's performance. Another paper [24] proposed an ST sliding-mode observer to improve the immunity of an FMSS system, and applied the conventional ST-SMC to design a voltage outer-loop controller to enhance its dynamic performance and robustness. However, the inner loop was poorly controlled using three-vector MPC, which requires high system data accuracy and complex computation.

In this study, a linear radiation model of a three-terminal FMSS was first obtained by applying the exact feedback linearization method, which decouples active and reactive currents. Subsequently, an adaptive QST-SMC (AQST-SMC) algorithm was proposed to design the voltage outer-loop and current inner-loop controllers of the three-terminal FMSS. The algorithm adopted the sigmoid function in the quasi-sliding mode to discretize the switching function and reduce the jitter. An adaptive scaling term was introduced to adjust the response speed of the system, which considerably shortens its response time and enhances immunity to interference. Finally, a three-terminal FMSS simulation model was constructed to verify the effectiveness of the proposed AQST-SMC strategy.

The rest of the paper is organized as follows: Section 2 describes the establishment of an affine nonlinear model of the FMSS system, which provides the basis for the subsequent construction of an AQST-SMC controller. Section 3 explains the construction of the AQST-SMC controller and compares it with conventional control strategies, such as first-order SMC. Section 4 describes the construction of the voltage outer-loop and current inner-loop controllers based on a radiative nonlinear model and mathematical equations of the FMSS system. Section 5 presents the establishment of a simulation of the three-terminal FMSS system, and the verification of its performance under different operating conditions. Section 6 presents the conclusions of this study.

2. FMSS Nonlinear Control Design

2.1. FMSS System Architecture

In medium-voltage distribution networks, feeders of different voltage levels are connected through an FMSS. However, with the upgrade and transformation of these distribution networks and the rapid development of loads, double-ended flexible switchgear cannot fulfill the requirements of the interconnection of multi-distribution areas in complex environments. In addition, the switchgear cannot regulate the power between the feeders if one of the ports fails. Therefore, it is important to study three-port and multiport FMSSs. A three-port FMSS can connect multiple feeders and reduce the construction cost. In addition, if one end is out of operation owing to failure, the rest of the converter can flexibly switch the working mode to support the normal operation of the power grid, which is more reliable. Figure 1 shows the structure of the three-terminal FMSS system mentioned earlier. For each feeder, the head of the transmission line is typically connected to the substation. Thus, the AC system connected to the FMSS can be equated to a Davignon equivalent circuit comprising a fixed alternating current (AC) voltage source and a series impedance at the point of common coupling.

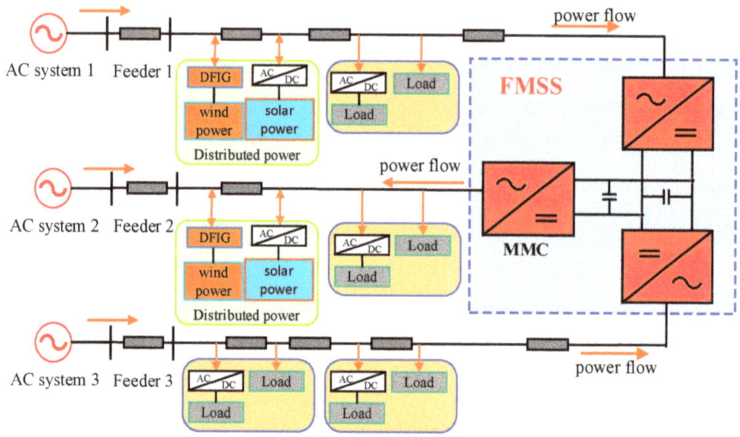

Figure 1. Structure diagram of three-terminal FMSS system.

Modular multilevel converters (MMCs) are widely used in power electronics, owing to their small loss and good quality of voltage and current waveforms generated by power conversion in the working state [25]. Thus, this study used MMCs as the main circuit topology.

2.2. FMSS Radiation Modeling

Assuming the AC system is balanced in three phases, the MMC equation of state was obtained from the circuit law as follows:

$$L\frac{d}{dt}\begin{bmatrix} i_{sd} \\ i_{sq} \end{bmatrix} = \begin{bmatrix} -R & \omega L \\ -\omega L & -R \end{bmatrix}\begin{bmatrix} i_{sd} \\ i_{sq} \end{bmatrix} + \begin{bmatrix} u_{sd} & -m_d U_{dc} \\ u_{sq} & -m_q U_{dc} \end{bmatrix}, \tag{1}$$

where ω represents the AC voltage angular frequency. In the rotating coordinate system, i_{sd} i_{sq} and u_{sd} u_{sq} represent the output current and voltage of any phase of the port, respectively. Moreover, m_d m_q represents the modulation switching function components of any phase of the port.

The transformation of (1) yielded the following nonlinear radiation model of the FMSS system:

$$\begin{cases} \dot{\boldsymbol{x}} = \boldsymbol{f}(\boldsymbol{x}) + \boldsymbol{g}(\boldsymbol{x})\boldsymbol{u} \\ \boldsymbol{y} = \boldsymbol{h}(\boldsymbol{x}) \end{cases}, \tag{2}$$

$$\boldsymbol{f}(\boldsymbol{x}) = \begin{bmatrix} f_1(\boldsymbol{x}) \\ f_2(\boldsymbol{x}) \end{bmatrix} = \begin{bmatrix} -\frac{R}{L}x_1 + \omega x_2 + \frac{u_{sd}}{L} \\ -\frac{R}{L}x_2 - \omega x_1 + \frac{u_{sq}}{L} \end{bmatrix}, \tag{3}$$

$$\boldsymbol{g}(\boldsymbol{x}) = \begin{bmatrix} g_1(\boldsymbol{x}) \\ g_2(\boldsymbol{x}) \end{bmatrix}^{\mathrm{T}} = \begin{bmatrix} -\frac{U_{dc}}{L} & 0 \\ 0 & -\frac{U_{dc}}{L} \end{bmatrix}, \tag{4}$$

where $\boldsymbol{x} = [x_1, x_2]^T = [i_{sd}, i_{sq}]^T$, $\boldsymbol{u} = [u_1, u_2]^T = [m_d, m_q]^T$, and $\boldsymbol{y} = [y_1, y_2]^T$. $\boldsymbol{h}(\boldsymbol{x})$ satisfies $\boldsymbol{h}(\boldsymbol{x}) = [\boldsymbol{h}_1(\boldsymbol{x}), \boldsymbol{h}_2(\boldsymbol{x})]^T = [i_{sdref-}x_1, i_{sqref-}x_2]^{\boldsymbol{T}}$, and i_{sdref}, i_{sqref} denotes the inner-loop current reference.

The Lie derivative of function $\boldsymbol{h}(\boldsymbol{x})$ along the vector field $\boldsymbol{f}(\boldsymbol{x})$ is defined as follows:

$$L_{f(x)}h_i(x) = \frac{\partial h_i(x)}{\partial x}f(x). \tag{5}$$

From (5), we obtain the following:

$$L_{g_i(x)}L_{f(x)}h_i(x) = \frac{\partial L_{f(x)}h_i(x)}{\partial x}g_i(x). \tag{6}$$

According to the differential geometry theory, the nonlinear radiation model satisfies the following constraints, and the matrix, $\boldsymbol{B}(\boldsymbol{x})$, is nonsingular:

$$\begin{cases} L_{f(x)}^{k}h_i(x) = L_{f(x)}\left(L_{f(x)}^{k-1}h_i(x)\right) \\ L_{g_i(x)}L_{f(x)}^{k}h_j(x) = 0 \\ L_{f(x)}^{0}h_i(x) = h_i(x) \\ L_{g_i(x)}L_{f(x)}^{r_j-1}h_j(x) \neq 0 \end{cases} \quad k = 1, 2, \cdots i, j = 1, 2. \tag{7}$$

$$\boldsymbol{B}(\boldsymbol{x}) = \begin{bmatrix} L_{g_1(x)}L_{f(x)}^{r_1-1}h_1(x) & L_{g_2(x)}L_{f(x)}^{r_1-1}h_1(x) \\ L_{g_1(x)}L_{f(x)}^{r_2-1}h_2(x) & L_{g_2(x)}L_{f(x)}^{r_2-1}h_2(x) \end{bmatrix} = \begin{bmatrix} \frac{U_{dc}}{L} & 0 \\ 0 & \frac{U_{dc}}{L} \end{bmatrix}, \tag{8}$$

where $r = [r_1, r_2]$ corresponds to the subrelativity of function $\boldsymbol{h}(\boldsymbol{x})$.

When the FMSS satisfies the normal operation of the aforementioned equation and $r_1 + r_2 = 2$ is equal to the order of the system, the choice of coordinate transformation can fully linearize the nonlinear model of the FMSS.

2.3. Coordinate Mapping and Control Law Solving

The following coordinate transformation matrix was selected:

$$z = \begin{bmatrix} z_1 \\ z_2 \end{bmatrix} = \begin{bmatrix} h_1(\boldsymbol{x}) \\ h_2(\boldsymbol{x}) \end{bmatrix} = \begin{bmatrix} i_{sdref} - x_1 \\ i_{sqref} - x_2 \end{bmatrix}. \tag{9}$$

The control variable after the exact feedback linearization takes the following form:

$$v = \boldsymbol{A}(\boldsymbol{x}) + \boldsymbol{B}(\boldsymbol{x})\boldsymbol{u} + \frac{R}{L}\boldsymbol{z}. \tag{10}$$

$$\boldsymbol{A}(\boldsymbol{x}) = \begin{bmatrix} L_{f(x)}^{r_1-1}h_1(x) \\ L_{f(x)}^{r_2-1}h_2(x) \end{bmatrix} = \begin{bmatrix} \frac{R}{L}x_1 - \omega x_2 - \frac{u_{sd}}{L} \\ \frac{R}{L}x_2 + \omega x_1 - \frac{u_{sq}}{L} \end{bmatrix}. \tag{11}$$

The reintegrated linear system was as follows:

$$\boldsymbol{u} = \boldsymbol{B}^{-1}(\boldsymbol{x})\left[v - \boldsymbol{A}(\boldsymbol{x}) - \frac{R}{L}\boldsymbol{z}\right]. \tag{12}$$

$$\boldsymbol{u} = \begin{bmatrix} m_d \\ m_q \end{bmatrix} = \begin{bmatrix} \frac{L}{U_{dc}}\left(v_1 + \omega x_2 + \frac{u_{sd}}{L}\right) - \frac{R}{U_{dc}}i_{sdref} \\ \frac{L}{U_{dc}}\left(v_2 - \omega x_1 + \frac{u_{sq}}{L}\right) - \frac{R}{U_{dc}}i_{sqref} \end{bmatrix}. \tag{13}$$

3. Construction of Adaptive Quasi-Super-Twisting Sliding Mode Controller

Second-order SMC uses the differentiation of the control input as a new virtual control quantity, and applies it to the higher-order derivatives of the sliding mode surface to ensure that the sliding variable, s, converges to zero in a restricted time [26]. This weakens the jitter, eliminates relative-order limitations, and improves the accuracy and reliability of the SMC.

The super-twisting algorithm (STA) is a second-order SMC algorithm designed for systems of relative order 1, which ensures that the system reaches the ideal state (s). Its first-order derivative is zero in a finite time. Conversely, conventional first-order SMC can only render the system state (s) zero in a finite time. The general expression of the STA is as follows:

$$\begin{cases} u = -\alpha|s|^{1/2}\mathrm{sign}(s) + u_1 \\ \dot{u}_1 = -\beta\mathrm{sign}(s) + \varphi \end{cases}, \tag{14}$$

where s represents a sliding variable; u_1 represents an auxiliary variable; φ represents a perturbation term; and α and β represent sliding mode gain coefficients.

If (14) satisfies $|\dot{\varphi}| \leq \delta$ and $\delta > 0$, then α and β simultaneously satisfying the following conditions can cause (14) to converge in a finite time [27].

$$\begin{cases} \alpha > 2 \\ \beta > \frac{\alpha^3+(4\alpha-8)\delta^2}{\alpha(4\alpha-8)} \end{cases} . \tag{15}$$

Figure 2 shows the phase plane trajectory of the STA, and point $S_{1-\mathrm{A}}$ is the intersection of the system-state trajectory curve with the axis, $\dot{s} = 0$ and $2(K'_{1-A}\beta - L)s_{1-A} = \dot{s}_0^2$, where K'_{1-A} and L are positive constants. The calculation shows that $\dot{s}_{1-A} = -\frac{2}{\alpha}\left(\frac{L}{K'_{1-A}} + \beta\right)s_{1-A}^{1/2}$, which affords $|\dot{s}_{1-A}|/|\dot{s}_0| < 1$. Considering that $|\dot{s}_1| \leq |\dot{s}_{1-A}|$, $|\dot{s}_1|/|\dot{s}_0| < 1$ holds; repeating the process ensures the inequality $|\dot{s}_{k+1}/\dot{s}_k| = |\dot{s}_{k+1}|/|\dot{s}_k| < 1$ holds. Similarly, it can be shown that $|s_{k+1}/s_k| = |s_{k+1}|/|s_k| < 1$ holds, indicating that the closed-loop system can converge to stability in a finite time [28].

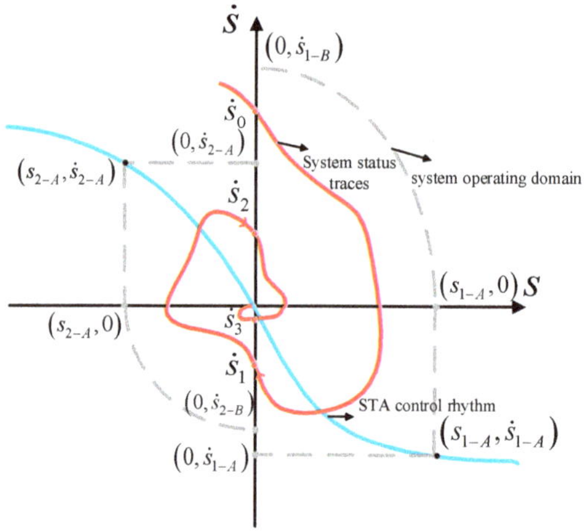

Figure 2. STA phase plane trajectory.

Based on (14), the introduction of the integral term can effectively attenuate the jitter and eliminate the static error of the system. However, there is still a sign function with poor smoothing performance near the sliding mode surface in the control law. This can cause high-frequency jitter with rapid and frequent switching on the sliding mode surface, and reduce the control accuracy of the system. A study [29] added a sinusoidal saturation function, sat (s), in the super-twisting control law to reduce the system jitter. However, the saturation function presents the problem of selecting the appropriate thickness of the boundary layer, which is extremely large to lead to the insufficient control accuracy of the system, and excessively small to cause oscillations. Thus, the following smoother sigmoid function was designed:

$$\mathrm{sd}(s) = \frac{2}{1+\mathrm{e}^{-ns}} - 1. \tag{16}$$

A constant ($n > 0$) was used to adjust the rate of the curve rise, which determines the function boundary layer thickness. Figure 3 shows that the sign function changes stepwise at $t = 0$, and that the sigmoid function continuously and smoothly changes with the input. The quasi-smooth mode changes more gradually than the switching function, which can alleviate the shock of the smooth mode switching and reduce the appearance of high-frequency jitter to a certain extent. This can improve the response speed and control accuracy of the FMSS system.

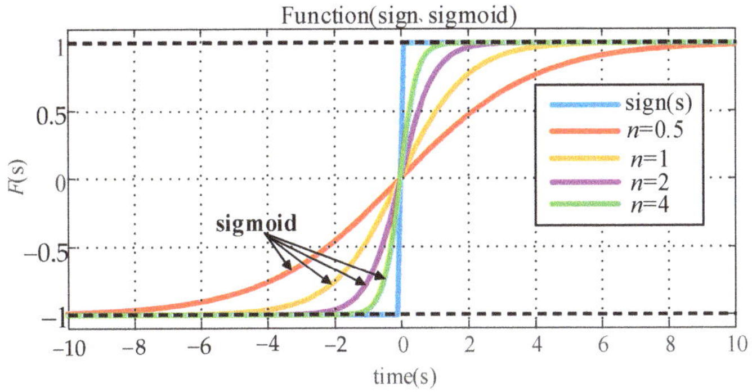

Figure 3. Symbolic and sigmoid functions for different values of n.

Equation (14) shows that the proportional term, $-\alpha|s|^{1/2}sign(s)$, improves the convergence speed of the algorithm. However, the exponent of the state vector is always 1/2, which prevents the system state from converging to a smaller region. The immunity of the algorithm to interference and the arrival speed are directly affected by its gain. A previous study [18] introduced the proportional term in a traditional STA to improve the convergence trajectory of the system. However, when $s < 1$, i.e., ks is zero; the proportional term is only left with the original square root term, and the problem of low arrival speed remains. To solve the aforementioned problem, this study introduced an adaptive linear term to realize the adaptive regulation of the convergence speed with the following expression:

$$q(s) = k|s|^{c\cdot \mathrm{sign}(|s|-1)}s, \tag{17}$$

where $k > 0$, $0 < c < 1$. When the system state tends to the sliding mode surface, i.e., $s < 1$, the adaptive term becomes $q(s) = k|s|^{-c}s$, $\alpha|s|^{1/2}\mathrm{sign}(s) + k|s|^{-c}s \gg \alpha|s|^{1/2}\mathrm{sign}(s)$, which speeds up the convergence. Furthermore, when the system state tends to the sliding mode surface, i.e., $s > 1$, the adaptive term becomes $q(s) = k|s|^{c}s$, $\alpha|s|^{1/2}\mathrm{sign}(s) + k|s|^{c}s \gg \alpha|s|^{1/2}\mathrm{sign}(s)$, which speeds up the convergence.

Using the sigmoid function and adaptive scaling term, the AQST-SMC mathematical expression can be written as follows:

$$\begin{cases} u = -\alpha|s|^{1/2}\mathrm{sd}(s) - q(s) + \omega \\ \dot{\omega} = -\beta \mathrm{sd}(s) + \dot{\varphi} \\ q(s) = k|s|^{c\cdot \mathrm{sign}(|s|-1)}s \end{cases}. \tag{18}$$

If (18) satisfies $2|\dot{\varphi}| \leq \delta$ and $\sigma > 0$, then α, β, and k simultaneously satisfying condition (19) can cause (18) to converge to the origin in a finite time. The stability of the AQST-SMC algorithm was confirmed, as described in Appendix A. It was discovered that the stability of the system will not be affected by the introduction of the adaptive linear term, $q(s)$, into the super-twisting algorithm, as long as it is guaranteed that $k > 0$.

$$\begin{cases} \alpha > 2 \\ \beta > \frac{\left(\alpha+k|x_1|^{1/2}\right)^2}{4\left(\alpha+k|x_1|^{1/2}\right)^2-8} + \frac{\delta^2}{4\alpha} \\ k > 0 \end{cases}. \tag{19}$$

To verify the effectiveness of the proposed AQST-SMC, it was compared with the conventional first-order SMC, conventional STA, and STA without the adaptive acceleration term. Figures 4 and 5 show the comparison results. First-order SMCs can usually reach the sliding mold surface in finite time. Nevertheless, they cannot maintain stability and equilibrium, and most conventional first-order SMCs are less efficient than second-order SMCs in terms of control performance and the tracking error, as shown in Figure 4a,b; the tracking accuracy and convergence speed of conventional first-order SMCs are worse than those of the conventional ST-SMC and the ST-SMC that contains only the sigmoid function. Using the designed adaptive convergence term equation $q(s)$, the AQST-SMC can approach the surface of the sliding mode quickly, and the system's state variables can converge smoothly. From Figure 5, the introduction of the continuous smooth sigmoid function attenuates the system's jitter and enhances the system's perturbation resistance compared to the sign function, confirming the effectiveness of the AQST-SMC.

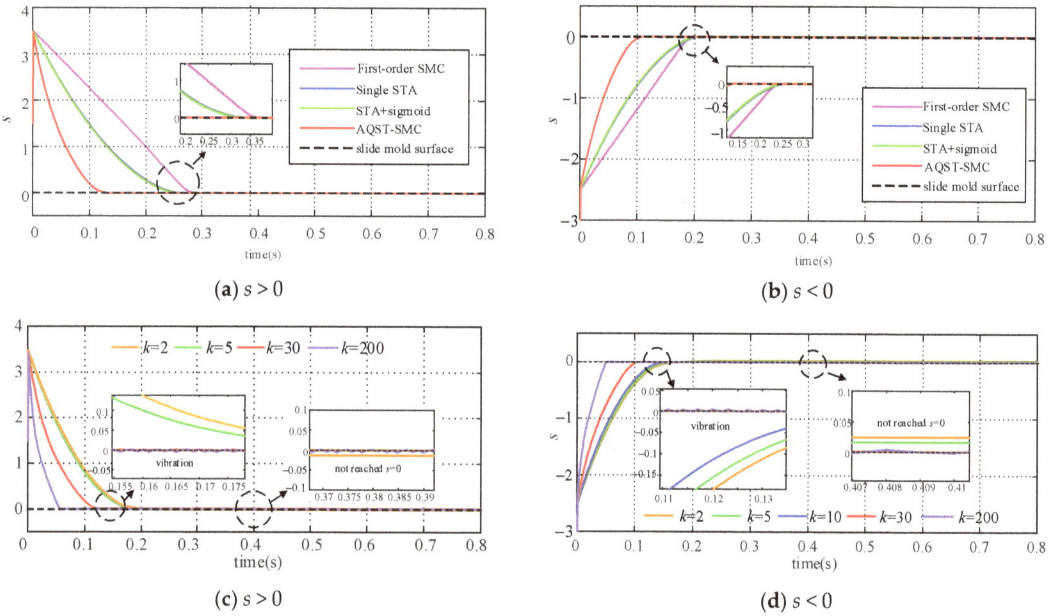

Figure 4. Comparison of the effect of different sliding mode control laws to converge to $s = 0$.

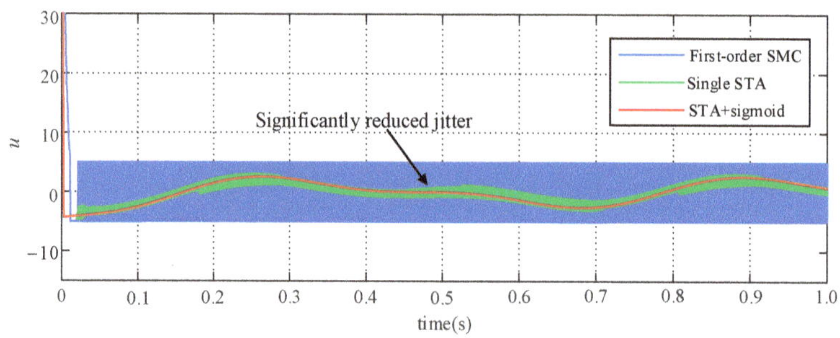

Figure 5. Comparison of vibration suppression effects of different sliding mode control laws.

Owing to the limitation of the stability condition Equation (19), control gain k must be selected within a certain range. When k is excessively small, the adaptive term $q(s)$ tends to zero, thus preventing the achievement of the desired acceleration effect and simultaneously departing from the limitations of the stability condition. This causes the system trajectory to deviate from the sliding mode surface; however, although the convergence speed increases when k becomes larger, it may cause new problems such as considerable overshooting or system oscillations due to the integral term. Therefore, k must be designed in a reasonable range to ensure that the system state variables converge smoothly in a stable range. Figure 4c,d show simulations that were conducted to test the effect of different gain coefficients k on the convergence of the system state to the surface of the sliding mode. When a higher value of k is considered (e.g., $k = 200$), the convergence is fast; thus, oscillations may occur. Conversely, when a small value of k is considered (e.g., $k = 2$ or 5), the system state does not reach the surface of the sliding mode completely, but the convergence speed is still faster than that of the conventional ST-SMC. Notably, in Figure 4, the three ST-SMCs use the same gain coefficients. Conversely, traditional first-order SMCs use the integral sliding mode surface and the exponential convergence law. Their gain parameters are selected to ensure that the system reaches the surface of the sliding mode in a finite amount of time and remains stable.

4. FMSS AQST-SMC Controller Design

4.1. AQST-SMC Voltage Outer Ring Design

During normal operation, the FMSS needs to operate in the $U_{dc}Q$ mode at one end to maintain a constant voltage on the DC side. The remaining ports operate in the PQ mode for power exchange. However, in the traditional outer-loop PI controller, when the system parameters are changed, the dynamic performance and response time of the system will increase because of the constant parameters of the PI. In this research, the AQST-SMC was designed to effectively suppress the jitter and speed up the response; it exhibits good dynamic performance and response time. The steady-state inverse model [7] was used to improve the traditional power outer-loop controller, which improves the outer-loop regulation rate and eliminates the static error of the system. For example, this study analyzed MMC1 running in the $U_{dc}Q$ operation mode, and MMC2 and MMC3 running in the PQ mode.

First, the three-port FMSS DC side satisfied (20), as follows:

$$\begin{cases} C\frac{dU_{dc}}{dt} = i_{dc1} + i_{dc2} + i_{dc3} = \sum\limits_{k=1}^{3} i_{dck} \\ i_{dck} = m_{ak}i_{ak} + m_{bk}i_{bk} + m_{ck}i_{ck} = \sum\limits_{j=a,b,c} m_{jk}i_{jk} \end{cases}, \tag{20}$$

where i_{dck} represents the DC side current of port k, and m_{jk}, i_{jk} denotes the one-phase switching function and output current of port k.

Combining (20) with coordinate mapping transformation shows the following:

$$\begin{aligned} C\frac{dU_{dc}}{dt} &= m_{ak}i_{ak} + m_{bk}i_{bk} + m_{ck}i_{ck} \\ &= \sum_{k=1}^{3} \frac{3}{2}\left(-m_{dk}i_{sdk} + m_{qk}i_{sqk}\right) \end{aligned}. \tag{21}$$

In the rotated coordinate system, m_{dk} and m_{qk} denote the port k switching function components. Moreover, u_{sdk} and i_{sdk} denote the voltage and current of port k, respectively.

$$\begin{cases} i_{sqk} = 0 \\ \frac{di_{sdk}}{dt} = 0 \end{cases}. \tag{22}$$

Equation (21) can be transformed into the following:

$$C\frac{dU_{dc}}{dt} = -\frac{3}{2}m_{d1}i_{sd1} - \frac{3}{2}m_{d2}i_{sd2} - \frac{3}{2}m_{d3}i_{sd3}, \tag{23}$$

where m_{dk} satisfies $m_{dk} = (\mathrm{u_{sdk}} - \mathrm{R_k i_{sdk}})/U_{dc}$.

The sliding mode surface s_0 was selected as the DC voltage tracking error.

$$s_0 = U_{dcref} - U_{dc}. \tag{24}$$

Equation (24) was differentiated as follows:

$$\dot{s}_0 = -\dot{U}_{dc}. \tag{25}$$

The uncertain disturbance φ was unknown; however, it was assumed to be bounded and satisfies $|\varphi| \leq m$, where m represents a known positive constant. To accurately track the reference DC voltage, it was obtained according to the AQST-SMC theory:

$$\begin{aligned} \dot{s}_0 &= -\dot{U}_{dc} = \alpha_0 |s_0|^{1/2} \mathrm{sd}(s_0) + k_0 |s_0|^{c_0 \cdot \mathrm{sign}(|s_0|-1)} s_0 + \int \beta_0 \mathrm{sd}(s_0) \mathrm{d}t \\ &= \frac{3}{2C}\left[\frac{u_{sd1} - R_1 i_{sd1}}{U_{dc}} \cdot i_{sd1} + \frac{u_{sd2} - R_2 i_{sd2}}{U_{dc}} \cdot i_{sd2} + \frac{u_{sd3} - R_3 i_{sd3}}{U_{dc}} \cdot i_{sd3}\right] \end{aligned}. \tag{26}$$

When the FMSS system is three-phase-balanced and in a steady state, the *d-axis* current reference value is obtained via simplification as follows:

$$\begin{aligned} i_{sdref} = &\frac{\frac{2C}{3} U_{dc}\left(\alpha_0 |s_0|^{1/2} \mathrm{sd}(s_0) + k_0 |s_0|^{c_0 \cdot \mathrm{sign}(|s_0|-1)} s_0 + \int \beta_0 \mathrm{sd}(s_0) \mathrm{d}t\right)}{u_{sd1} - R_1 i_{d1}} \\ &+ \frac{-i_{sd2}(u_{sd2} - R_2 i_{sd2}) - i_{sd3}(u_{sd3} - R_3 i_{sd3})}{u_{sd1} - R_1 i_{sd1}} \end{aligned}. \tag{27}$$

Figure 6 shows the AQST-SMC-based FMSS voltage outer-loop control.

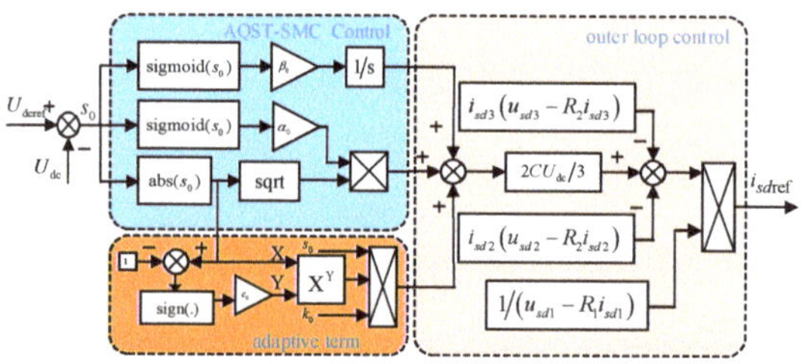

Figure 6. Voltage outer loop AQST-SMC block diagram.

4.2. AQST-SMC Current Inner Loop Design

The current inner-loop sliding mode surfaces s_1 and s_2 of the FMSS were expressed as follows:

$$s = \begin{bmatrix} s_1 \\ s_2 \end{bmatrix} = \begin{bmatrix} z_1 \\ z_2 \end{bmatrix}. \tag{28}$$

When the system reaches a steady state, the derivatives of s_1 and s_2 are zero, and the system control law is as follows:

$$v = \begin{bmatrix} v_1 \\ v_2 \end{bmatrix} = \begin{bmatrix} \frac{R}{L} z_1 - \alpha_1 |s_1|^{1/2} \mathrm{sd}(s_1) - k_1 |s_1|^{c_1 \cdot \mathrm{sign}(|s_1|-1)} s_1 - \beta_1 \mathrm{sd}(s_1) \\ \frac{R}{L} z_2 - \alpha_2 |s_2|^{1/2} \mathrm{sd}(s_2) - k_2 |s_2|^{c_2 \cdot \mathrm{sign}(|s_2|-1)} s_2 - \beta_2 \mathrm{sd}(s_2) \end{bmatrix}. \tag{29}$$

Furthermore, the original system control volume is obtained as the following:

$$u = \frac{L}{U_{dc}}\begin{bmatrix} v_1 \\ v_2 \end{bmatrix} + \frac{1}{U_{dc}}\begin{bmatrix} -Ri_{sdref} + \omega L x_2 + u_{sd} \\ -Ri_{sqref} - \omega L x_1 + u_{sq} \end{bmatrix}. \tag{30}$$

Figure 7 shows a schematic of the final FMSS AQST-SMC integrated control strategy.

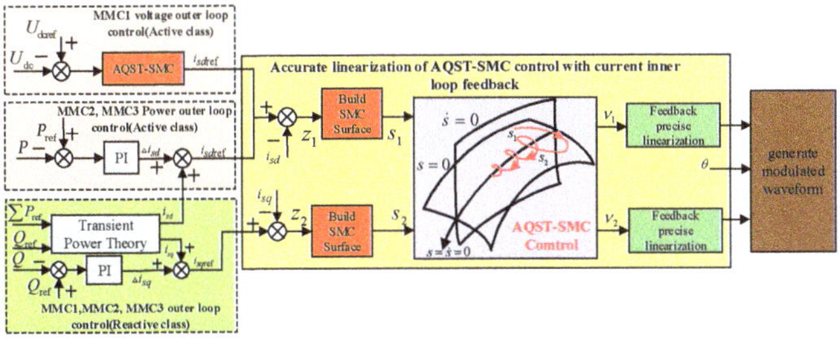

Figure 7. Block diagram of three-terminal FMSS AQST-SMC control strategy.

5. Analysis of System Simulation Examples

According to the demand of a distribution network under different working conditions, the FMSS mainly operates in the $U_{dc}Q$, PQ, and $U_{ac}f$ modes. To enable the FMSS to precisely control the current and fast power supply restoration, the following collaborative control methods are adopted [30]:

(1) During normal operation, the MMC of one port typically operates in the $U_{dc}Q$ mode to stabilize the DC bus voltage and regulate the reactive power of that port. The MMCs of the rest of the ports operate in the PQ mode to separately regulate the active and reactive power of the ports.

(2) The MMC running in the $U_{dc}Q$ mode is the more vulnerable link in the entire system. When the MMC running in this mode fails, it is necessary to switch the MMC in the PQ mode to the $U_{dc}Q$ mode to maintain the DC side voltage.

(3) In the event of a port feeder failure, the FMSS can rapidly isolate the fault and switch to the $U_{ac}f$ mode to provide continuous power to critical loads in the area experiencing loss.

The three-port FMSS model (Figure 1) was built using MATLAB 2023a/Simulink software. Table 1 lists the system parameters of the three ports.

Table 1. Simulation parameters of the FMSS system.

System Parameter	Symbol	Value
AC voltage rating	u_s	10 kV
AC side-rated frequency	f_s	50 Hz
Rated port capacity	S	10 MVA
DC voltage reference	U_{dc}	40 kV
Submodule point capacitance	C	5000 μF
Bridge arm inductance	L_0	2 mH
Number of sub-modules	N	20

5.1. *AQST-SMC Performance Analysis*

The simulations of the conventional first-order SMC, conventional STA, and improved AQST-SMC were compared to study the optimization performance of the algorithms and validate the theoretical analyses presented in Section 3. The traditional first-order SMC uses an integral sliding mode surface and the exponential convergence law. In addition, the sign function in the original exponential convergence law is replaced by a saturation function, which is more effective in suppressing jitter (Appendix A). The Pref and Q_{ref} of port 2 were 0.2 p.u. and 0.3 p.u. At t = 0.02 s, the active and reactive powers of port 2 were assumed to suddenly increase and decrease by 0.2 p.u., respectively. Figure 8 shows the simulation results.

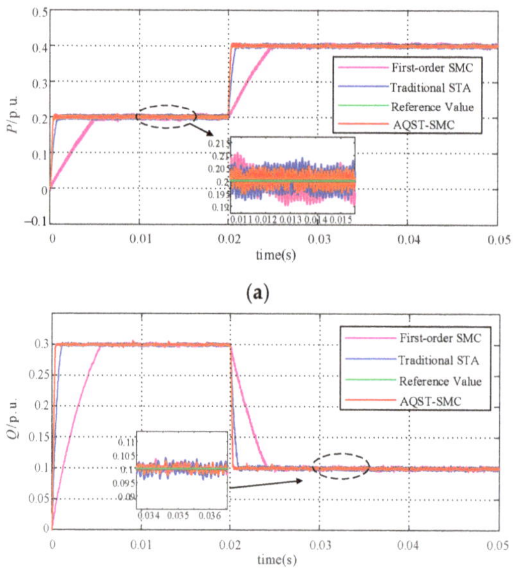

Figure 8. Power response characteristics of port 2. (**a**) Active and (**b**) reactive power characteristic curves.

The power response curves presented in Figure 8a,b show that the proposed AQST-SMC method had a faster response speed at the startup moment of the system. Furthermore, it reached a steady state in a shorter time because of the rapid adjustment of convergence when the power changed. Based on the calculations, the power oscillation of the AQST-SMC algorithm using the sigmoid function was considerably smaller in the steady-state operation than in other SMCs. It was approximately 66.71% and 60.13% of the active and reactive oscillations, respectively, under the proposed STA, compared with that under the traditional STA, effectively reducing the power oscillation in the steady-state operation and improving the transient performance of the system.

Table 2 lists the output power response time at the system startup moment and the regulation time to return to a steady state when disturbed. Table 2 shows that the active power response times at the startup moment using the proposed AQST-SMC method were 1.96 and 8.43 times those of the traditional STA and first-order SMC algorithm, respectively. The reactive power response times were 2.61 and 11.70 times better than those of the traditional STA and first-order SMC algorithm, respectively. The active power regulation time to return to a steady state when disturbed under the AQST-SMC algorithm improved compared with those of the traditional STA and first-order SMC algorithm. The active power regulation times of the AQST-SMC algorithm were 3.75 and 12.34 times better than those of the traditional STA and first-order SMC algorithm, respectively. Finally, the reactive power regulation times were 3.33 and 14.67 times better than those of the traditional STA and first-order SMC algorithm, respectively. Thus, the proposed AQST-SMC method effectively improved the speed of the system to reach a steady state, and considerably reduced the degree of fluctuation of the system under transient operation. In addition, the robustness and anti-interference capability of the system improved.

Table 2. Performance characteristics of FMSSs.

Control Modes	Response Time (ms)	Adjustment Time (ms)	Steady-State Oscillation Amplitude
First-order SMC	P: 4.30 Q: 5.50	P: 3.95 Q: 4.40	P: 9.87% Q: 8.91%
Traditional STA	P: 1.00 Q: 1.23	P: 1.20 Q: 1.00	P: 6.96% Q: 8.36%
AQST-SMC	P: 0.51 Q: 0.47	P: 0.32 Q: 0.30	P: 4.64% Q: 5.03%

5.2. *Dynamic Simulation Verification of FMSS System*

To further investigate the dynamic performance and anti-interference ability of the control system, this study compared the simulations of several typical working conditions prone to occur under new energy access. The initial operating state of the three-port FMSS is as follows: the output power value of port 1 is Q_{1ref} = 0 p.u.; the output power values of port 2 are P_{2ref} = 0.1 p.u. and Q_{2ref} = 0.3 p.u.; and the output power values of port 3 are P_{3ref} = 0.2 p.u. and Q_{3ref} = 0.5 p.u. Table 3 lists the parameters of different controllers. The first-order traditional SMC features an additional control action based on the exponential convergence law. Notably, the present study only applied the AQST-SMC in port MMC1 (traditional $U_{dc}Q$ outer loop) and MMC2, and 3 (traditional PQ control outer loop) used PI control. Furthermore, the controller parameters were calculated and then carefully selected for the optimal steady-state operation of the system, as well as to achieve the desired response speed and convergence.

Table 3. Parameters of different FMSS controllers.

Control Method	Parameters of Outer Loop Controller	Parameters of Inner Loop Controller
Traditional PI (MMC 1)	$k_{p1} = 0.5$, $k_{i1} = 100$	$k_{p1} = 22$, $k_{i1} = 3.46$
Traditional PI (MMC 2)	$k_{p2} = 0.00005$, $k_{i2} = 70$	$k_{p2} = 22$, $k_{i2} = 3.46$
Traditional PI (MMC 3)	$k_{p3} = 0.000028$, $k_{i3} = 80$	$k_{p3} = 22$, $k_{i3} = 3.46$
First-Order SMC	$\varepsilon = 1$, $q = 40$, $c_1 = 30$, $\Delta = 0.05$	$\varepsilon = 3$, $q = 4000$, $c_1 = 30$, $\Delta = 0.05$
AQST-SMC	$\alpha_1 = 2$, $\beta_1 = 30$, $k_1 = 10$	$\alpha_1 = 110{,}000$, $\beta_1 = 20{,}000$, $k_1 = 10$

5.2.1. Comparison of Simulations with Disturbed System Output Power

The active power variation affects the stability of the DC-side voltage. The simulation of port 3 was subjected to power fluctuation after accessing distributed power supply and a new type of load. The traditional PI and proposed AQST-SMC algorithm were simulated and compared; Figure 9 shows the output feedback curves of the three-terminal FMSS.

Figure 9 shows that the system startup moment, using the traditional PI control under the output power response, was slow. It took approximately 0.1 s to stabilize. Based on the active power reference of MMC3 at t = 0.2 s, the voltage jumped from 0.2 (standard) to 0.5 to simulate the power fluctuation of the FMSS system caused by the distributed power supply and new load access. Under PI control, the system required approximately 0.12 s to reach a steady state after being perturbed. The dynamic mathematical model of the DC side of the three-port FMSS and the active power balance relationship showed that the active power perturbation affected the stability of the voltage on the DC side [9]; moreover, owing to the inability of the PI controller to adapt to the dynamics of the system under the sudden change of power, the DC voltage fluctuated considerably at 0.2 s and 0.45 s, and the sizes of the oscillations were approximately 112.05 V and 105.85 V, respectively. When the reactive power of MMC1 fluctuated at 0.35 and 0.55 s, its active power suffered from a certain degree of fluctuation, verifying the proposed control strategy for the effectiveness of the power decoupling. Under the proposed AQST-SMC, the system stabilized within

0.001 s at the startup moment and after being disturbed. The dynamic response time was reduced by 100 times, and the DC voltage remained stable. Thus, the feasibility of the AQST-SMC was verified, the anti-disturbance performance of the system considerably improved, the decoupling between the output powers of the system was realized, and the stability of the additive system improved.

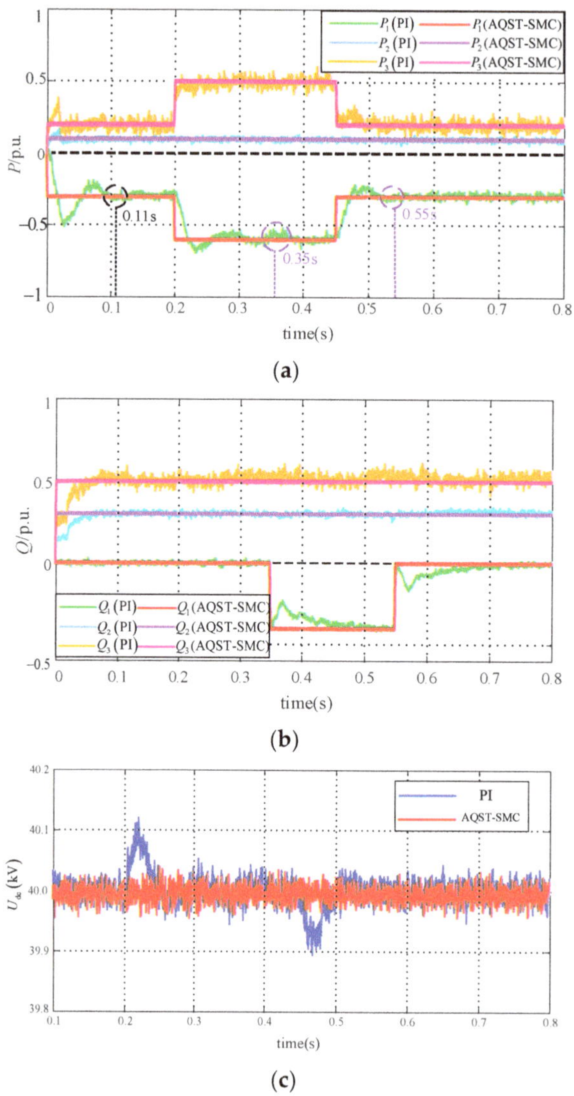

Figure 9. Output feedback of FMSS system under power perturbations. (**a**) Active power, (**b**) reactive power, and (**c**) DC voltage response curves.

5.2.2. System AC Measured Voltage Amplitude Dips

To validate the proposed method for AC fault measurements, the MMC3 AC voltage was simulated at an amplitude drop of −0.4 (units) in 0.2–0.3 s. Figure 10a shows the AC voltage and current responses; the current response was rapid and accurate. Figure 10b,c show that during voltage dips, the proposed AQST-SMC exhibited good transient characteristics at instances of fault occurrences and good steady-state characteristics during the fault period, compared with the conventional first-order SMC and conventional PI control. The active and reactive powers were rapidly maintained, with fast response and robustness. Figure 11a,b show that during voltage dips at the MMC3 side, MMC2 can function normally without being affected, and its active and reactive powers can remain stable. This indicated that the dynamic response of the system was fast, with good anti-interference properties.

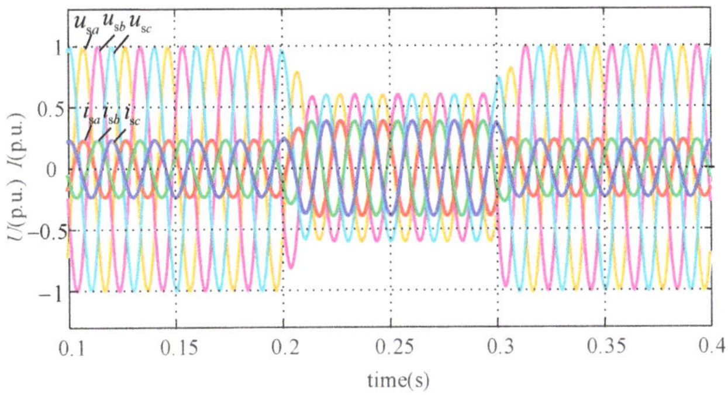

(**a**)

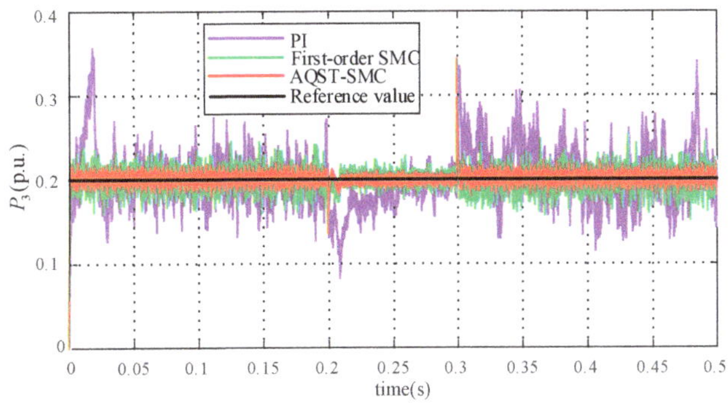

(**b**)

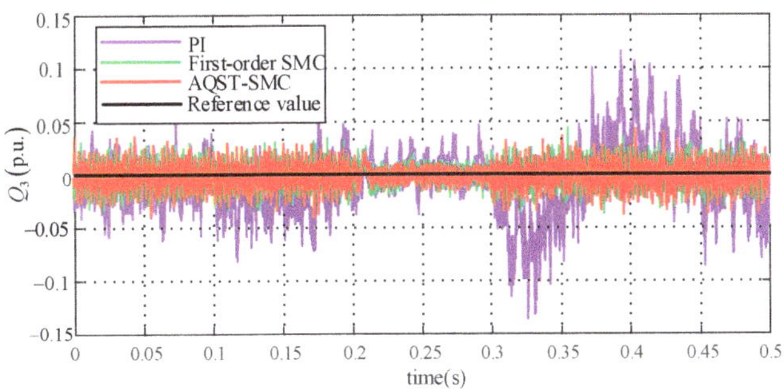

(**c**)

Figure 10. MMC3 output feedback under voltage drop. (**a**) MMC3 voltage and current. (**b**) MMC3 active power. (**c**) MMC3 reactive power.

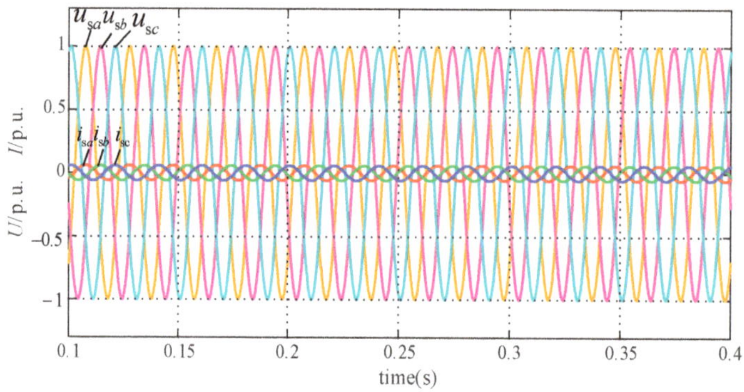

(a)

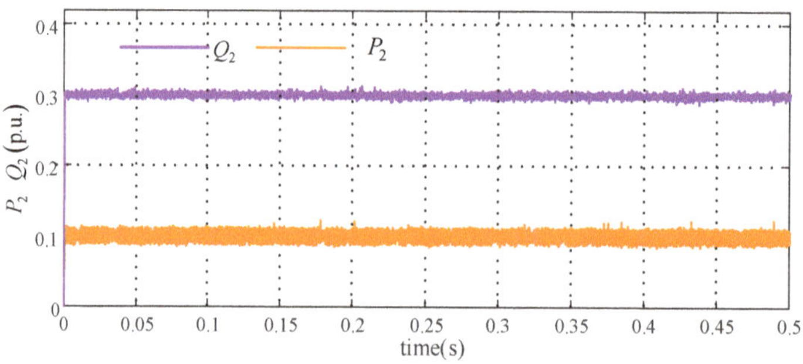

(b)

Figure 11. MMC2 output feedback under voltage drop. (**a**) MMC2 voltage and current. (**b**) MMC2 active and reactive power.

5.2.3. Disturbance of System Electrical Parameters

During practical operation, the AC-side equivalent impedance of the system generated parameter perturbations in response to changes in grid conditions, environmental factors, and load conditions. To analyze the immunity of the system, it was assumed that the AC inductance on the MMC1 side increased to thrice its original value and reduced to 40% of the original value at t = 0.2 s. Figures 12 and 13 show the response curves of the d-axis currents and DC side voltages using the three methods. Under PI control, the system exhibited a slow response at the initial moment, and produced fluctuations in the d-axis currents of approximately 93 A when the inductance parameters were changed. However, the STA and proposed AQST-SMC overcame the problem of slow response at the initial startup moment of the system. The ability of the proposed AQST-SMC to track the reference value of the d-axis current in the steady state and after parameter changes was superior to that of the conventional STA, and the jitter considerably reduced. The DC voltage of the proposed AQST-SMC responded rapidly and fluctuated to a lesser extent at the time of parameter ingestion. Furthermore, the voltage was stabilized at the reference value of 40 kV, confirming the superiority of the AQST-SMC strategy.

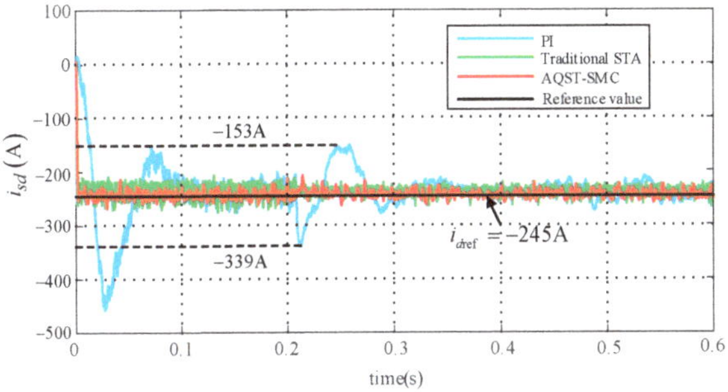

(a)

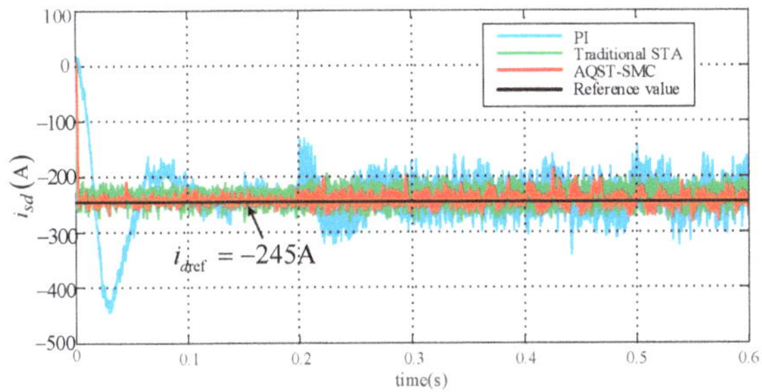

(b)

Figure 12. Current response of phase a of MMC1 under parameter perturbation. (**a**) L tripled. (**b**) L reduced to 40% of the original value.

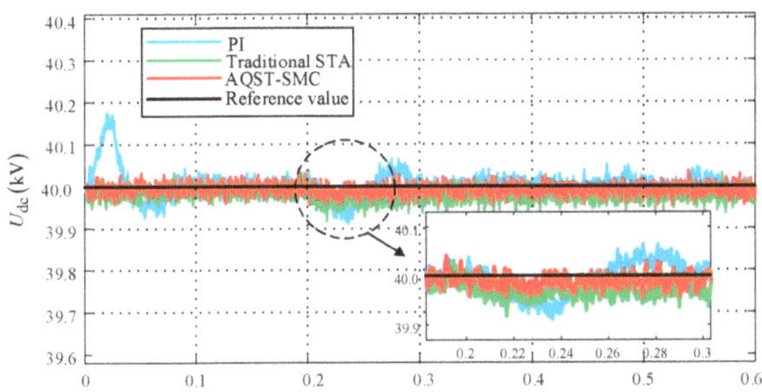

(a)

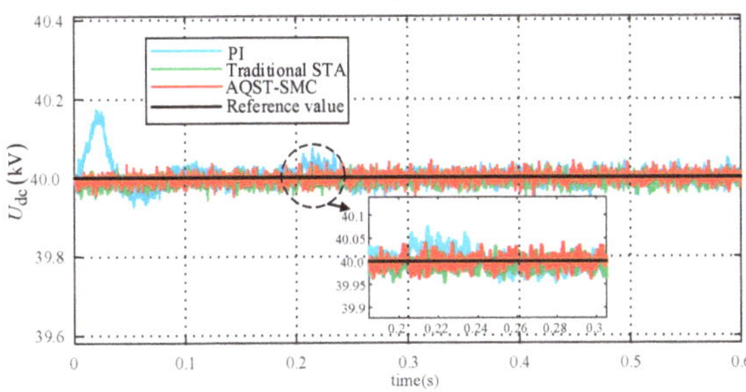

(b)

Figure 13. DC voltage response under parameter perturbation. (**a**) L tripled. (**b**) L reduced to 40% of the original value.

5.3. Validation of $U_{ac}f$ Mode of Operation

When the feeder connected to the FMSS fails, to guarantee the operation of important loads, a stable AC voltage and grid frequency must be provided to the power loss area; this is called the $U_{ac}f$ mode of operation. To verify the effectiveness of the AQST-SMC method during power loss, port 2 was set to fail and the load connected at the end was purely resistive (value = 120). Figure 14 shows the three-port AC measurement output response. Figure 14a shows the output voltage and current waveforms of the failed port 2. The output voltage and current were smooth and stable, and the faulty power supply function was realized. Figure 14b,c show that the voltage and current waveforms of ports 1 and 3 remained stable and unaffected by port 2. This confirmed the feasibility of the proposed control strategy, and ensured continuous power supply in the faulty area.

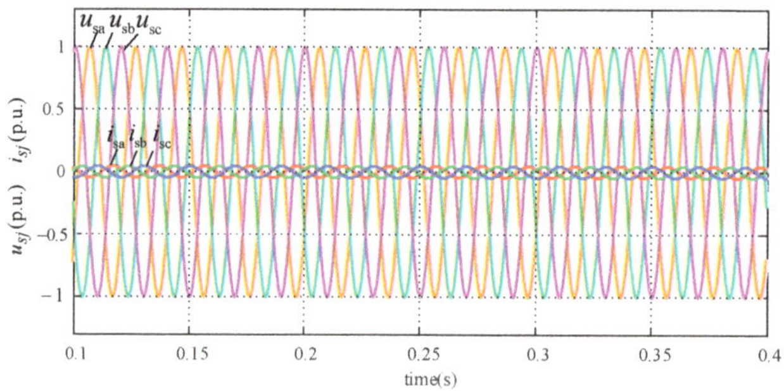

(**a**)

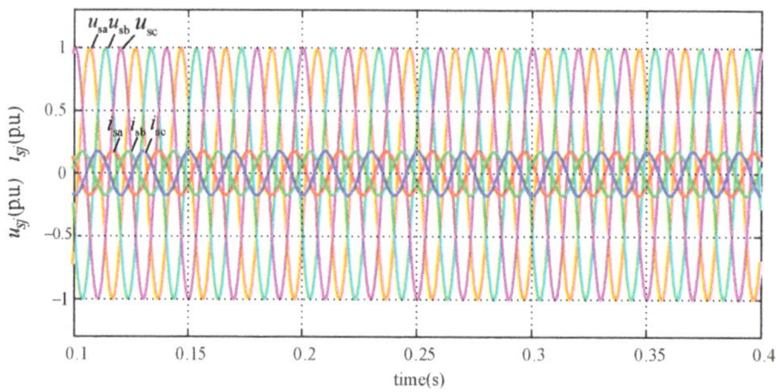

(**b**)

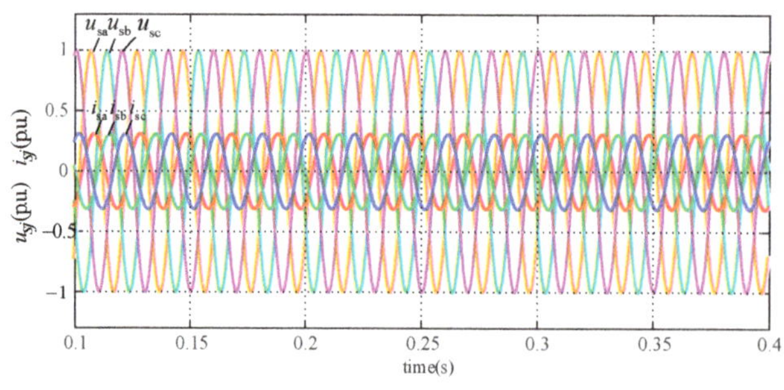

(**c**)

Figure 14. AC side response of ports in the $U_{ac}f$ mode. Voltage and current of ports (**a**) 2, (**b**) 1, and (**c**) 3.

6. Conclusions

This study proposes adaptive super-twisting SMC based on feedback-accurate linearization decoupling for the nonlinear and strongly coupled mathematical model of an FMSS in the distribution network and the difficulty in the traditional PI double-closed-loop control parameter tuning. The main conclusions are as follows:

(1) The three-terminal FMSS nonlinear radiation model was transformed into a linear model via exact feedback linearization and applied to the FMSS inner-loop control design. This completely decoupled the output power, achieved a simple controller design, and improved the operating characteristics of the system.

(2) To address the large jitter vibration and low convergence speed of the first-order SMC, STA control was introduced. AQST-SMC was designed using the sigmoid function and adaptive proportional term to realize better jitter vibration suppression and higher convergence speed, respectively. The inner- and outer-loop composite control system of the FMSS was designed to effectively improve the dynamic response speed, robustness, and steady-state characteristics of the system.

(3) The proposed method improved the transient and steady-state characteristics of the system, with good anti-interference ability under several working conditions with high incidence rates. This can provide insights into innovative solutions for the application of FMSSs in smart distribution networks and new power systems.

Although this study implemented the AQST-SMC algorithm in the $U_{dc}Q$ and PQ operating modes with the corresponding validation analyses, the dynamic process of switching between different operating modes has not yet been considered. In the future, the different operating conditions of FMSSs should be investigated from the perspective of mode switching. In addition, the proposed AQST-SMC should be applied to the virtual synchronous control of new grid-type energy sources to improve its robustness and response speed.

Author Contributions: W.M. and X.W.: methodology, software, validation, formal analysis, writing—original draft preparation, supervision, and writing—review and editing; Y.W.: project administration and funding acquisition; W.Z., H.L. and Y.Z.: investigation, resources, and data curation. All authors have read and agreed to the published version of the manuscript.

Funding: This research was funded by the National Natural Science Foundation of China, grant number 52277208.

Data Availability Statement: Data are contained within the article.

Acknowledgments: The authors would like to express their gratitude to all those who helped them during the writing of this paper. The authors would like to thank the reviewers for their valuable comments and suggestions.

Conflicts of Interest: The authors declare no conflicts of interest.

Appendix A

For the actual situation of the system, the constant-value disturbance can be a special case of the time-varying disturbance [31]. Simultaneously, the adaptive term evolved from the proportional term, *ks*; thus, the quadratic-like Lyapunov function was used to perform the STA stability proofs in terms of the time-varying perturbation.

First, (17) was replaced using the following variables:

$$\begin{cases} x_1 = s \\ \dot{x}_1 = u = -\alpha|x_1|^{1/2}\mathrm{sd}(x_1) - kx_1 + x_2 \\ x_2 = \omega \\ \dot{x}_2 = -\beta\mathrm{sd}(x_1) + \dot{\varphi} \end{cases}. \tag{A1}$$

The following class of quadratic Lyapunov functions [32] was selected:

$$V(x_1, x_2) = \zeta^{\mathrm{T}} P\zeta, \tag{A2}$$

where $\zeta^{\mathrm{T}} = \begin{bmatrix}\zeta_1 & \zeta_2\end{bmatrix} = \begin{bmatrix}|x_1|^{1/2}\mathrm{sd}(x_1) & x_2\end{bmatrix}$; real symmetric positive definite matrices P and ζ satisfy the following:

$$P = \begin{bmatrix} 4\beta + \left(\alpha + k|x_1|^{1/2}\right)^2 & -\left(\alpha + k|x_1|^{1/2}\right) \\ -\left(\alpha + k|x_1|^{1/2}\right) & 2 \end{bmatrix}, \tag{A3}$$

$$\begin{aligned} \dot{\zeta} &= \begin{bmatrix} \frac{1}{2}|x_1|^{-\frac{1}{2}}\left(-\alpha|x_1|^{\frac{1}{2}}\mathrm{sd}(x_1) - kx_1 + x_2\right) \\ -\beta\mathrm{sd}(x_1) + \dot{\varphi} \end{bmatrix}, \\ &= \frac{1}{2|\zeta_1|}(A\zeta + B\phi) \end{aligned} \tag{A4}$$

where $A = \begin{bmatrix} -\left(\alpha + k|x_1|^{1/2}\right) & 1 \\ -2\beta & 0 \end{bmatrix}$, $B = \begin{bmatrix} 0 \\ 1 \end{bmatrix}$, $\phi = 2|\zeta_1|\dot{\varphi}$.

According to (17), $V(x_1, x_2)$ is a radially unbounded continuous positive definite function, and $V(x_1, x_2)$ is differentiable to any point except the set $\{x_1 = x_2 = 0\}$, for which the following derivation was obtained:

$$\begin{aligned} \dot{V}(x_1, x_2) &= \dot{\zeta}^{\mathrm{T}} P\zeta + \zeta^{\mathrm{T}} P\dot{\zeta} \\ &= \frac{1}{2|\zeta_1|}\left[(A\zeta)^{\mathrm{T}} + (B\phi)^{\mathrm{T}}\right] P\zeta + \zeta^{\mathrm{T}} P \frac{1}{2|\zeta_1|}(A\zeta + B\phi) \\ &= \frac{1}{2|\zeta_1|}\left[\zeta^{\mathrm{T}} A^{\mathrm{T}} P\zeta + \phi^{\mathrm{T}} B^{\mathrm{T}} P\zeta + \zeta^{\mathrm{T}} PA\zeta + \zeta^{\mathrm{T}} PB\phi\right] \\ &= \frac{1}{2|\zeta_1|}\begin{bmatrix} \zeta \\ \phi \end{bmatrix}^{\mathrm{T}} \begin{bmatrix} A^{\mathrm{T}}P + PA & PB \\ B^{\mathrm{T}}P & 0 \end{bmatrix} \begin{bmatrix} \zeta \\ \phi \end{bmatrix} \end{aligned} \tag{A5}$$

ϕ is a scalar, and an expansion of $B^{\mathrm{T}}P\zeta$ reveals that it is also a scalar; therefore, we obtain the following:

$$B^{\mathrm{T}}P\zeta = \left(B^{\mathrm{T}}P\zeta\right)^{\mathrm{T}} = \zeta^{\mathrm{T}}P^{\mathrm{T}}\left(B^{\mathrm{T}}\right)^{\mathrm{T}} = \zeta^{\mathrm{T}}PB. \tag{A6}$$

Taking $n = B^{\mathrm{T}}P\zeta = \zeta^{\mathrm{T}}PB$ and $n^2 = \zeta^{\mathrm{T}}PBB^{\mathrm{T}}P\zeta$, this follows from the inequality shown below:

$$(n - \phi)^2 = n^2 - 2n\phi + \phi^2 \geq 0, \forall n, \phi \in R. \tag{A7}$$

Through (A7), the following results:

$$\zeta^{\mathrm{T}}PBB^{\mathrm{T}}P\zeta \geq \phi^{\mathrm{T}}B^{\mathrm{T}}P\zeta + \zeta^{\mathrm{T}}PB\phi - \phi^2. \tag{A8}$$

By (19), the perturbation term was assumed to be Lipschitz continuous $2|\dot{\varphi}| \leq \delta$ and $\delta > 0$.

Combined with $\phi = 2|\zeta_1|\dot{\varphi}$, this can be obtained by simplifying the following inequality operation:

$$\delta^2\zeta_1^2 - \phi^2 \geq 0. \tag{A9}$$

At $C = [1\ 0]$, the following can be obtained:

$$\delta^2\zeta_1^2 = \delta^2\zeta^{\mathrm{T}}C^{\mathrm{T}}C\zeta. \tag{A10}$$

Combining (A5) through (A10), the following conclusions were drawn:

$$\begin{aligned}\dot{V}(x_1,x_2) &= \frac{1}{2|\zeta_1|}\begin{bmatrix}\zeta\\ \phi\end{bmatrix}^{\mathrm{T}}\begin{bmatrix}A^{\mathrm{T}}P+PA & PB\\ B^{\mathrm{T}}P & 0\end{bmatrix}\begin{bmatrix}\zeta\\ \phi\end{bmatrix}\\ &\le \frac{1}{2|\zeta_1|}\left\{\begin{bmatrix}\zeta\\ \phi\end{bmatrix}^{\mathrm{T}}\begin{bmatrix}A^{\mathrm{T}}P+PA & PB\\ B^{\mathrm{T}}P & 0\end{bmatrix}\begin{bmatrix}\zeta\\ \phi\end{bmatrix}+\delta^2\zeta_1^2-\phi^2\right\}\\ &\le \frac{1}{2|\zeta_1|}\zeta^{\mathrm{T}}\left[A^{\mathrm{T}}P+PA+\delta^2C^{\mathrm{T}}C+PBB^{\mathrm{T}}P\right]\zeta\\ &= -\frac{1}{2|\zeta_1|}\zeta^{\mathrm{T}}Q\zeta\end{aligned}. \quad \text{(A11)}$$

Let $\left[A^{\mathrm{T}}P+PA+\delta^2C^{\mathrm{T}}C+PBB^{\mathrm{T}}P\right]=-Q<0$; therefore, according to Equation (A11), we obtain the following:

$$\dot{V}(x_1,x_2)=-\frac{1}{2|\zeta_1|}\zeta^{\mathrm{T}}Q\zeta<0. \quad \text{(A12)}$$

Equation (A12) shows that the Lyapunov function of the system satisfies the stability conditions in the Lyapunov stability theory, which is radially unbounded, and a < b. However, when a time-varying perturbation exists, it is not guaranteed that a symmetric positive definite matrix P can be found to satisfy the algebraic equations for any positive definite matrix, Q. However, the idea of (15) can be used to find the range of values of the control parameter that ensures the finite time convergence of the system, such that Q is positive definite.

By solving the calculation for the Q matrix, the following expansion was obtained:

$$\begin{aligned}Q &= -\left[A^{\mathrm{T}}P+PA+\delta^2C^{\mathrm{T}}C+PBB^{\mathrm{T}}P\right]\\ &= \begin{bmatrix}2\left(\alpha+k|x_1|^{1/2}\right)^3-\left(\alpha+k|x_1|^{1/2}\right)^2+4\beta\left(\alpha+k|x_1|^{1/2}\right)-\delta^2 & 2\left(\alpha+k|x_1|^{1/2}\right)-2\left(\alpha+k|x_1|^{1/2}\right)^2\\ 2\left(\alpha+k|x_1|^{1/2}\right)-2\left(\alpha+k|x_1|^{1/2}\right)^2 & 2\left(\alpha+k|x_1|^{1/2}\right)-4\end{bmatrix}\end{aligned}. \quad \text{(A13)}$$

If Q is nonnegative, then $\dot{V}<0$; it can be deduced that a sufficient condition for Q to be a symmetric positive definite matrix is the following:

$$\begin{cases}\alpha+k|x_1|^{1/2}>2\\ \beta>\dfrac{\left(\alpha+k|x_1|^{1/2}\right)^3+\delta^2\left(\alpha+k|x_1|^{1/2}\right)-2\delta^2}{4\left(\alpha+k|x_1|^{1/2}\right)\left[\left(\alpha+k|x_1|^{1/2}\right)-2\right]}\end{cases}. \quad \text{(A14)}$$

Considering that $k>0$, and there must be $|x_1|^{1/2}\ge 0$ in the FMSS system, (A14) can be transformed as follows:

$$\begin{cases}\alpha>0\\ \beta>\dfrac{\left(\alpha+k|x_1|^{1/2}\right)^2}{4\left(\alpha+k|x_1|^{1/2}\right)-8}+\dfrac{\delta^2}{4\alpha}\\ k>0\end{cases}. \quad \text{(A15)}$$

According to Lyapunov's stability theorem, the system is stable when the conditions in (A15) are satisfied, and there is a symmetric positive definite matrix Q in that $\dot{V}<0$.

The traditional first-order SMC used in this paper is the exponential convergence law and the integral slip mode surface, the expression of which is shown below:

$$\begin{cases}\nu_1=U_{\mathrm{dcref}}-U_{\mathrm{dc}}\\ s_1=\nu_1+c_1\int \nu_1 dt\\ \dot{s}=-\varepsilon sat(s_1)-qs_1 \; \varepsilon,q>0\end{cases}. \quad \text{(A16)}$$

where s_1 represents the sliding-mode surface, c_1 is the sliding-mode surface control parameter, ν_1 is the DC voltage error signal of the FMSS system, and ε and q are the exponential convergence law control parameters. Saturation function $sat(s_1)$ was selected for reducing the system jitter.

References

1. Chen, Y.; Tang, Z.; Weng, X.; He, M.; Zhang, G.; Yuan, D.; Jin, T. A novel approach for evaluating power quality in distributed power distribution networks using AHP and S-transform. *Energies* **2024**, *17*, 411. [CrossRef]
2. Liu, Y.; Hu, J.; Shi, J.; Yang, B. Adaptive integrated control strategy for MMC-MTDC transmission system considering dynamic frequency response and power sharing. *J. Electr. Eng. Technol.* **2023**, *147*, 108858. [CrossRef]
3. Ivic, D.R.; Stefanov, P.C. An extended control strategy for weakly meshed distribution networks with soft open points and distributed generation. *IEEE Access* **2021**, *9*, 137886–137901. [CrossRef]
4. Mardanimajd, K.; Karimi, S.; Anvari-Moghaddam, A. Voltage stability improvement in distribution networks by using soft open points. *J. Electr. Eng. Technol.* **2024**, *155*, 109582. [CrossRef]
5. Li, Z.; Ye, Y.; Wang, Z.; Wu, Y.; Xu, H. Integrated planning and operation evaluation of micro-distribution network based on flexible multistate switch interconnection. *CES Trans. Electr. Mach. Syst.* **2021**, *36*, 10. [CrossRef]
6. Hou, Y.; Xu, Y.; Wang, Z.; Chen, X.; Cui, H. Research on application of three-port SNOP based on dual closed-loop control in the distribution network. In Proceedings of the IEEE International Conference on Smart Technologies and Management for Computing, Chennai, India, 2–4 August 2017; pp. 389–395. [CrossRef]
7. Wang, Z.; Sheng, L.; Huo, Q.; Hao, S. An improved model predictive control method for three-port soft open point. *Math. Probl. Eng.* **2021**, *2021*, 9910451. [CrossRef]
8. Peng, B.; Zhang, G. Coordination control strategy for three-port SNOP based on FCS-MPC. *J. Eng.* **2019**, *2019*, 1005–1010. [CrossRef]
9. Aithal, A.; Li, G.; Wu, J.; Yu, J. Performance of an electrical distribution network with soft open point during a grid side AC fault. *Appl. Energy* **2018**, *227*, 262–272. [CrossRef]
10. Li, Z.; Hao, Q.; Gao, F.; Wu, L.; Guan, M. Nonlinear decoupling control of two-terminal MMC-HVDC based on feedback linearization. *IEEE Trans. Power Deliv.* **2019**, *34*, 376–386. [CrossRef]
11. Zhang, G.; Tang, B.; Shen, C.; Wang, T.; Xu, C.; Xia, Z. A strategy for suppression of voltage fluctuation on DC side when SOP port is unbalanced. *Electr. Meas. Instrum.* **2021**, *61*, 95–101. [CrossRef]
12. Li, B.; Liang, Y.; Wang, G.; Li, H.; Ding, J. A control strategy for soft open points based on adaptive voltage droop outer-loop control and sliding mode inner-loop control with feedback linearization. *Int. J. Electr. Power Energy Syst.* **2020**, *122*, 106205. [CrossRef]
13. Zhou, M.; Su, H.; Zhou, H.; Feng, Y.; Wang, D.; Cheng, J. Full-order sliding mode control strategy for soft open point. *Proc. CSEE* **2023**, *43*, 8622–8635. [CrossRef]
14. Liu, Y.C.; Laghrouche, S.; Depernet, D.; N'Diaye, A.; Djerdir, A.; Cirrincione, M. Disturbance-observer-based speed control for SPMSM drives using modified super-twisting algorithm and extended state observer. *Asian J. Control* **2024**, *26*, 1089–1102. [CrossRef]
15. Xu, Q.; Jiang, D.; Wang, Y.; Zhang, X.; Liu, J.; Chen, Y. Variable-step close-loop angle compensation method of PMSM rotor position estimation based on super-twisting sliding-mode observer using tangent reaching law. *Energy Rep.* **2023**, *9*, 356–361. [CrossRef]
16. Maged, N.A.; Hasanien, H.M.; Ebrahim, E.A.; Tostado-Véliz, M.; Jurado, F. Optimal super twisting sliding mode control strategy for performance improvement of islanded microgrids: Validation and real-time study. *Int. J. Electr. Power Energy Syst.* **2024**, *157*, 109849. [CrossRef]
17. Ghazi, G.A.; Al-Ammar, E.A.; Hasanien, H.M.; Ko, W.; Lee, S.M.; Turky, R.A.; Tostado-Véliz, M.; Jurado, F. Circle search algorithm-based super twisting sliding mode control for MPPT of different commercial PV modules. *IEEE Access* **2024**, *12*, 33109–33128. [CrossRef]
18. Saadaoui, A.; Ouassaid, M. Super-twisting sliding mode control approach for battery electric vehicles ultra-fast charger based on Vienna rectifier and three-phase interleaved DC/DC buck converter. *J. Energy Storage* **2024**, *84*, 110854. [CrossRef]
19. Çelik, D.; Ahmed, H.; Meral, M.E. Kalman filter-based super-twisting sliding mode control of shunt active power filter for electric vehicle charging station applications. *IEEE Trans. Power Deliv.* **2023**, *38*, 1097–1107. [CrossRef]
20. Tiwary, N.; Naik, N.V.; Panda, A.K.; Narendra, A.; Lenka, R.K. A robust voltage control of DAB converter with super-twisting sliding mode approach. *IEEE J. Emerg. Sel. Top. Ind. Electron.* **2023**, *4*, 288–298. [CrossRef]
21. Wang, F.; Yang, A.; Yu, X.; Zhang, Z.; Wang, G. Model-free predictive current control for three-phase Vienna rectifier based on adaptive super-twisting sliding mode observer. *Trans. China Electrotech. Soc.* **2024**, *39*, 1859–1870. [CrossRef]
22. Pati, A.K.; Sahoo, N.C. Adaptive super-twisting sliding mode control for a three-phase single-stage grid-connected differential boost inverter based photovoltaic system. *ISA Trans.* **2017**, *69*, 296–306. [CrossRef]
23. Hou, J.; Wang, Z.; Zhou, H.; Xu, H.; Wei, W. Soft open point control strategy based on super-twisting-full-order fast terminal composite sliding mode. *Autom. Instrum.* **2023**, *3*, 143–147. [CrossRef]
24. Wang, Z.; Zhou, H.; Su, H. Disturbance observer-based model predictive super-twisting control for soft open point. *Energies* **2022**, *15*, 3657. [CrossRef]

25. Ma, W.; Lü, Q.; Zhang, Y.; Zhang, K.; Han, J.; Li, M. Suppression strategy for fault current of flexible multi-state switch in distribution network fault state. *Power Syst. Technol.* **2021**, *45*, 4251–4258. [CrossRef]
26. Zhang, K.; Wang, L.; Fang, X.; Zhang, K.; Han, J.; Li, M. High-order fast nonsingular terminal sliding mode control of permanent magnet linear motor based on double disturbance observer. *IEEE Trans. Ind. Appl.* **2022**, *58*, 3696–3705. [CrossRef]
27. Li, P. Convergence of super-twisting algorithm based on quadratic-like Lyapunov function. *Control Decis.* **2011**, *26*, 949–952. [CrossRef]
28. Levant, A. Sliding order and sliding accuracy in sliding mode control. *Int. J. Control* **1993**, *58*, 1247–1263. [CrossRef]
29. Yang, H.; Bai, Y.; Chien, Y.R. Generalized super-twisting sliding mode control of permanent magnet synchronous motor based on sinusoidal saturation function. *IEICE Electron. Express* **2022**, *19*, 20220066. [CrossRef]
30. Fu, K.; Yi, H.; Zhuo, F.; Li, S. Self-healing control strategy based on snop under severe fault condition. In Proceedings of the 2021 IEEE Sustainable Power and Energy Conference (iSPEC), Nanjing, China, 23–25 December 2021. [CrossRef]
31. Li, X.; Liu, J.; Yin, Y.; Zhao, K. Improved super-twisting non-singular fast terminal sliding mode control of interior permanent magnet synchronous motor considering time-varying disturbance of the system. *IEEE Access* **2023**, *11*, 17485–17496. [CrossRef]
32. Tripathi, V.K.; Kamath, A.K.; Behera, L.; Verma, N.K.; Nahavandi, S. Finite-time super twisting sliding mode controller based on higher-order sliding mode observer for real-time trajectory tracking of a quadrotor. *IET Control Theor. Appl.* **2020**, *14*, 2359–2371. [CrossRef]

energies

MDPI

Article

Grid-Forming Control for Solar Generation System with Battery Energy Storage

Yupeng Cai [1], Lujie Yu [2,*], Meng Wu [1], Shengyang Lv [1], Ziyu Fu [2], Wenhao Tong [2], Wei Li [1] and Songjie Shi [1]

[1] State Grid Liaoning Electric Power Co., Ltd., Electric Power Science Research Institute, Shenyang 110006, China; cai8872@126.com (Y.C.); wumeng_ldk@163.com (M.W.); sgdlck@126.com (S.L.); yzylw2007@126.com (W.L.); shisongjie2009@126.com (S.S.)

[2] School of Electrical and Information Engineering, Tianjin University, Tianjin 300072, China; fuziyu2021@163.com (Z.F.); 15958966510@163.com (W.T.)

* Correspondence: lujie.yu@outlook.com

Abstract: Solar generation systems with battery energy storage have become a research hotspot in recent years. This paper proposes a grid-forming control for such a system. The inverter control consists of the inner dq-axis current control, the dq-axis voltage control, the phase-locked loop (PLL) based frequency control, and the DC voltage control. The proposed control embeds the PLL into the grid-forming inverter control, offering the advantages of better synchronization and fault current-limiting capability. With the proposed control, the battery energy storage is able to provide inertial and primary frequency support during the grid frequency disturbance. Simulation models are established in PSCAD/EMTDC, and the results during the active power variation and AC voltage variation, the grid frequency disturbance, grid fault, and mode switch validate the effectiveness of the proposed control.

Keywords: battery energy storage; frequency control; grid-forming control; inverter control; solar generation

1. Introduction

The use of fossil energy sources and air pollution issues drive the fast development of renewable energy. Among all the renewable energy sources, wind power generation and solar generation are the most attractive ones. In the US, wind generation accounts for 10% of the total electrical generation while solar generation will take up 5% of the total electrical generation by the end of 2023. China has set the goal that before 2030 carbon emissions will reach their peak value and before 2060 carbon neutrality will be achieved. By the first quarter of 2023, solar generation in China will exceed 228 GW while wind generation in China will exceed 310 GW. It has been reported that China is now the leader in renewable energy installation capacity worldwide and holds around 50% of the total wind and solar capacity.

However, the high penetration of renewable energy sources raises some concerns about power system stability. For the wind and solar energy sources, they are all connected to the power grid through power-electronic converters. These converters generally cannot provide the same inertial or damping as the traditional synchronous generators. In [1,2], the grid-connected converters for renewable generation sources are classified as grid-following converters and grid-forming converters.

Among the renewable generation sources in commission, they generally operate on the grid-following mode [3]. However, with the increased integration of the grid-following renewable generation sources and the decrease in the grid strength, the converter-driven oscillation becomes a serious and annoying issue. In [4], a small signal state space model is established for VSC-HVDC, and eigenvalue results show that the maximum power generated is significantly affected by the PLL and the system finds it difficult to maintain stability when the short circuit ratio is smaller than 1.3. Based on the work in [4], the authors of [5,6] establish the small signal model for doubly fed induction generator (DFIG)-based wind turbines and fully rated converter-based wind turbines, finding that the PLL and rotor-side converter power control has a noticeable effect on the DFIG connected system stability, while the weak grid, high wind power export, low voltage, and low PLL bandwidth are the main causes for the stability issues in the latter system. In [7], the harmonic state–space model considering the PLL under the asymmetrical grid is established and it is observed that the stability region of the system is reduced during the increase in the impedance asymmetrical index. In [8,9], an impedance model is built, and a generalized Nyquist plot is used to determine the system stability. It is observed that the negative resistor characteristic of *qq* impedance, which is mainly caused by the PLL, reduces the system stability margin. In [10], an improved feedforward control method considering PLL dynamics is proposed to improve the stability of the grid-following converter connected to a weak grid. In [11], a current error-based compensation control is proposed, where a magnitude compensation is implemented with a proportional controller whose input is the reactive current, and an angle compensation is realized with a proportional and integral (PI) controller whose input is the active current. Both eigenvalues and simulation results show the above control increases the system stability, especially when the short circuit ratio of the system is small. In [12], another advanced vector control is proposed for the grid-following converter, where the error between the active power reference and the measured one is not only the input of the active power PI controller but also the input of the reactive power PI controller. Such a control improves the system stability but suffers from the multi-parameter tuning.

An alternative way to increase the system's small signal stability is to change the grid-following control to the grid-forming control. In [13,14], a power synchronization control is proposed using the internal synchronization mechanism in AC systems and tries to emulate a synchronous machine with the converter and the results show that the control enables the converter to provide a strong voltage support to the weak grid. In [15,16], the virtual synchronous generator and droop control-based grid-forming converter are compared, and the results show the virtual synchronous generator has larger inertia than the droop control, but when a proper lead–lag component is embedded in the active power droop control, the same small signal model has been derived with the virtual synchronous generator model. In [17,18], the sequence impedance model of the virtual synchronous generator is provided, and the stability analysis of the voltage-controlled virtual synchronous generator is compared to the current-controlled virtual synchronous generator. It is concluded that the voltage-controlled virtual synchronous generator is more similar to the traditional synchronous generator and is more stable under a weak grid when compared with the current-controlled virtual synchronous generator. In [19,20], the inertial support and primary frequency support from different types of the grid-forming converter are compared, and the simplified virtual synchronous compensator shows the advantage of providing frequency support and damping during the active power change and frequency variation.

In all, current research focuses on the control and stability analysis of the grid-forming converters without the PLL, and the synchronization is implemented through the power control during the power generation mode. But the reality is that the PLL still plays a key role for the grid-forming converter. The main reasons are as follows: (1) when the offline renewable generation sources try to connect to the grid, the phase of the grid needs to be known for smooth connection; (2) when a fault happens, the current limiting control needs a PLL. Thus, in this paper, a grid-forming control with a reserved PLL is proposed for the converters for solar generation.

The rest of this paper is organized as follows. Section 2 describes the topology of the solar generation. Section 3 presents the proposed grid-forming control for the solar generation converter. Section 4 presents the simulation under the power and voltage variation, during the grid frequency change, AC fault, and the mode switch from the grid-connected mode to the islanded mode. Finally, Section 5 draws the conclusion.

2. Topology of the Studied System

Figure 1 shows the studied grid-connected solar generation, which mainly consists of the following parts: (1) a photovoltaic (PV) array; (2) a front boost converter; (3) a battery energy storage system (BESS); and (4) an inverter. The PV array transforms the solar energy to the electricity, the DC power is injected to the front boost converter, the battery is connected to the DC bus through a bidirectional DC/DC converter, the DC power collected from both the PV and the BESS is then converted to AC power through the grid-connected inverter. The AC voltage at the inverter terminal is filtered by the LC circuit and is increased to a collection level through the transformer.

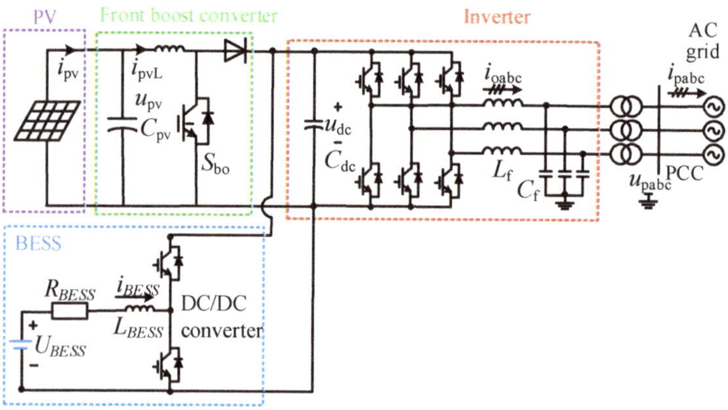

Figure 1. Topology of the studied system.

3. Proposed Grid-Forming Control for Solar Generation

This section presents the control of the solar generation, mainly includes the control of the front boost converter, control of the BESS, and the proposed grid-forming control of the inverter.

3.1. Control of the Front Boost Converter

The dynamics of the capacitor and the inductance of the PV front boost converter can be expressed as

$$\begin{cases} C_{pv}\frac{du_{pv}}{dt} = i_{pv} - i_{pvL} \\ L_{pv}\frac{di_{pvL}}{dt} = u_{pv} - (1 - d_{pv})u_{dc} \end{cases} \quad (1)$$

where C_{pv} and L_{pv} are the capacitor and inductance value, u_{pv} and i_{pv} are the voltage and current of the PV array, i_{pvL} is the inductance current, u_{dc} is the DC voltage of the grid-connected inverter, and d_{pv} is the duty cycle of the front boost converter. Based on the dynamics of the capacitor and the inductance of the PV front boost converter, the control of the front boost converter for the maximum power point tracking (MPPT) is shown in Figure 2 and can be expressed as

$$d_{pv} = \left(k_{pb} + \frac{k_{ib}}{s}\right)\left(u_{pv}^{*} - u_{pv}\right) \quad (2)$$

where u_{pv}^{*} is the voltage reference for the u_{pv}, which is obtained from the MPPT, and k_{pb} and k_{ib} are the proportional and integral parameters of the controller.

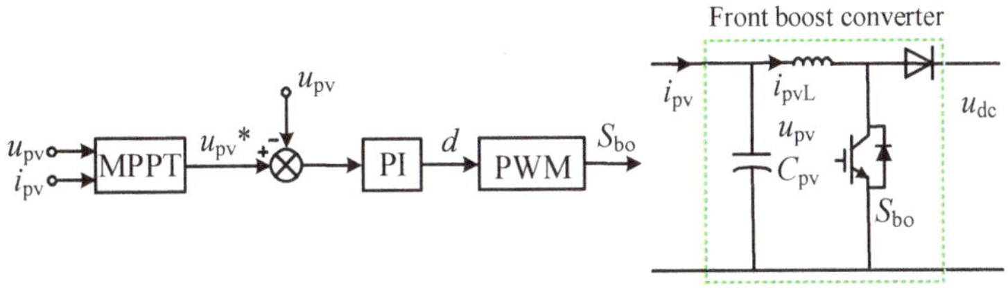

Figure 2. Control of the front boost converter.

3.2. Control of the BESS

For the BESS, the control structure consists of the two layers, as shown in Figure 3. The upper-level control takes the Deblock signal, state of charge (SOC), and discharge/charge mode as the input and outputs the signal Enboost, Enbuck, and $P_{refBESS}$. Figure 4 shows the flow chart of the control logic. "Enboost" works under the following conditions: (1) The Deblock signal is equal to 1, which indicates the DC/DC converter is not blocked. (2) The SOC is larger than 5%, which indicates the battery has enough stored energy to support the AC grid. (3) The discharge signal is equal to 1, which indicates the power needs to be transmitted from the battery to the grid and the DC/DC converter should operate on the boost control mode. On the other hand, "Enbuck" works under the following conditions: (1) The Deblock signal is equal to 1, which indicates the DC/DC converter is not blocked. (2) The SOC is less than 100%, which indicates the battery has the capacity to absorb the energy. (3) The charge signal is equal to 1, which indicates the power needs to be transmitted from the grid or solar generation to the battery and the DC/DC converter should operate on the buck control mode.

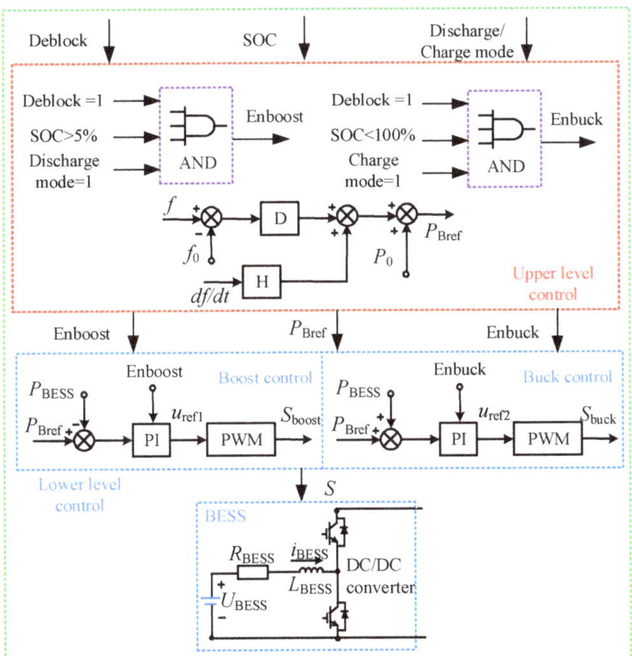

Figure 3. Control of the BESS.

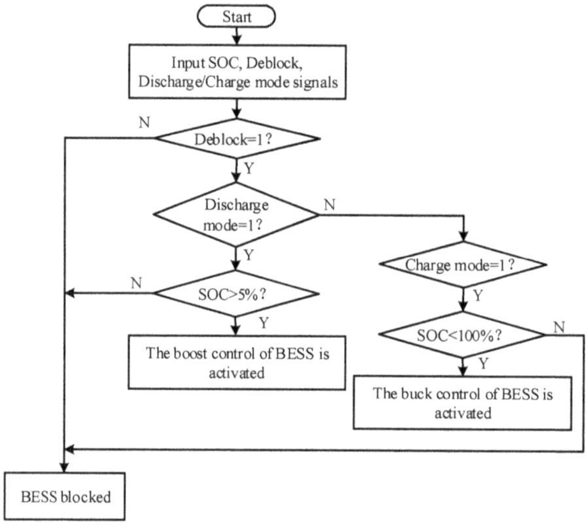

Figure 4. Flow chart of the control logic.

The dynamic of the resistor and the inductance can be expressed as

$$L_{\text{BESS}}\frac{\mathrm{d}i_{\text{BESS}}}{\mathrm{d}t} = u_{\text{BESS}} - u_{\text{t}} + R_{\text{BESS}}i_{\text{BESS}} \tag{3}$$

where L_{BESS} and R_{BESS} are the inductance and resistance of the BESS, u_{BESS} is the voltage of the battery, and i_{BESS} is the output current of the battery. The control for the bidirectional DC/DC converter in the BESS is expressed as

$$\begin{cases} u_{\text{ref1}} = (P_{\text{Bref}} - P_{\text{BESS}})(k_{\text{pdc1}} + \frac{k_{\text{idc1}}}{s}) \text{ Boost/discharging mode} \\ u_{\text{ref2}} = (P_{\text{Bref}} + P_{\text{BESS}})(k_{\text{pdc2}} + \frac{k_{\text{idc2}}}{s}) \text{ Buck/charging mode} \end{cases} \tag{4}$$

where P_{BESS} is the active power of the BESS, P_{Bref} is the active power reference of the BESS, k_{pdc1} and k_{idc1} are the proportional and integral parameters for the power control in the boost/discharging mode, k_{pdc2} and k_{idc2} are the proportional and integral parameters for the power control in the buck/charging mode, and u_{ref1} and u_{ref2} are the voltage references for the pulse modulation width (PMW) in the two different modes, respectively.

When the grid frequency changes, in order to support the system inertia and primary frequency, the BESS active power reference P_{Bref} is determined by the upper-level control as

$$P_{\text{Bref}} = H\frac{\mathrm{d}f}{\mathrm{d}t} + D(f - f_0) + P_0 \tag{5}$$

where H is the inertial time constant, D is the parameter coefficient for the primary frequency support, f is the grid frequency, f_0 is the nominal frequency, and P_0 is the active power feedforward term.

3.3. Grid-Forming Control of the Inverter

The inverter control mainly consists of the inner current control, inner voltage control, PLL-based frequency control and the DC voltage control, as shown in Figure 5.

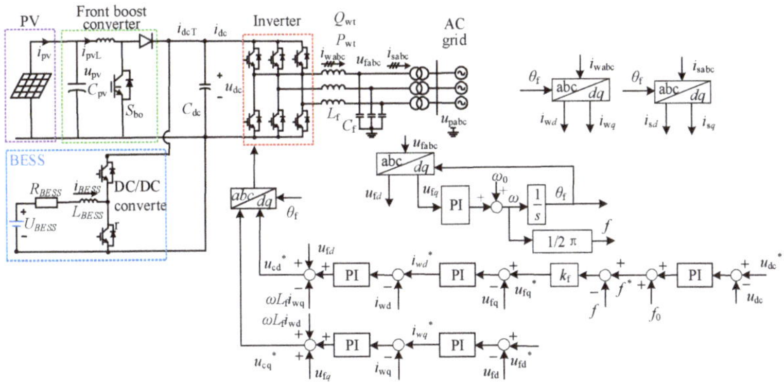

Figure 5. Proposed grid-forming control of the inverter.

For the grid-connected inverter, the dynamic of the inductance in the *dq* frame is expressed as

$$\begin{cases} L_{\mathrm{f}}\frac{di_{\mathrm{wd}}}{dt} = u_{\mathrm{cd}} - u_{\mathrm{fd}} + \omega L_{\mathrm{f}} i_{\mathrm{wq}} - R_{\mathrm{f}} i_{\mathrm{wd}} \\ L_{\mathrm{f}}\frac{di_{\mathrm{wq}}}{dt} = u_{\mathrm{cq}} - u_{\mathrm{fq}} - \omega L_{\mathrm{f}} i_{\mathrm{wd}} - R_{\mathrm{f}} i_{\mathrm{wq}} \end{cases} \tag{6}$$

where L_{f} and R_{f} are the inductance and resistance of the grid-connected converter, u_{cd} and u_{cq} are the *dq*-axis voltage at the inverter terminal, u_{fd} and u_{fq} are the *dq*-axis voltage at the inverter filter, i_{wd} and i_{wq} are the *dq*-axis converter current, and ω is grid angular frequency.

Thus, the inner current control is designed as

$$\begin{cases} u_{\mathrm{cd}}^{*} = \left(k_{\mathrm{pi}} + \frac{k_{\mathrm{ii}}}{s}\right)\left(i_{wd}^{*} - i_{\mathrm{wd}}\right) - \omega L_{\mathrm{c}} i_{\mathrm{wq}} + u_{\mathrm{fd}} \\ u_{\mathrm{cq}}^{*} = \left(k_{\mathrm{pi}} + \frac{k_{\mathrm{ii}}}{s}\right)\left(i_{\mathrm{wq}}^{*} - i_{\mathrm{wq}}\right) + \omega L_{\mathrm{c}} i_{\mathrm{wd}} + u_{\mathrm{fq}} \end{cases} \tag{7}$$

where u_{cd}^{*} and u_{cq}^{*} are the *dq*-axis inverter voltage references, i_{wd}^{*} and i_{wq}^{*} are the active and reactive current references, i_{wd} and i_{wq} are the *dq*-axis inverter current, k_{pi} and k_{ii} are the parameters of the inner current control.

On the other hand, when the inverter behaves like an AC voltage, the active power and the reactive power P_{inv} and Q_{inv} can be expressed as

$$\begin{cases} P_{\mathrm{inv}} = \frac{U_{\mathrm{f}} E \sin(\theta_{\mathrm{f}} - \theta_{\mathrm{e}})}{X_{\mathrm{ac}}} \\ Q_{\mathrm{inv}} = \frac{(U_{\mathrm{f}} - E\cos(\theta_{\mathrm{f}} - \theta_{\mathrm{e}}))U_{\mathrm{f}}}{X_{\mathrm{ac}}} \end{cases} \tag{8}$$

where U_{f} is the RMS value of the line-to-line voltage of the inverter, E is the RMS value of the line-to-line voltage of the grid, θ_{f} and θ_{f} are the phase angles of the voltages at the inverter filter and the AC grid, respectively, and X_{ac} is the inductance between the inverter and the grid.

Considering that the relationship between the active power and reactive power and the *dq*-axis inverter current is

$$\begin{cases} P_{\mathrm{inv}} = 1.5 u_{\mathrm{fd}} i_{\mathrm{wd}} + 1.5 u_{\mathrm{fq}} i_{\mathrm{wq}} \\ Q_{\mathrm{inv}} = -1.5 u_{\mathrm{fd}} i_{\mathrm{wq}} + 1.5 u_{\mathrm{fq}} i_{\mathrm{wd}} \end{cases} \tag{9}$$

and the θ_{f} is mainly determined by the u_{fq}, as shown in Figure 6, the *d*-axis current can be used to adjust the u_{fq}, while the *q*-axis current can be used to adjust the u_{fd}, thus the voltage control of the inverter is designed as

$$\begin{cases} i_{\mathrm{wd}}^{*} = \left(k_{\mathrm{pv1}} + \frac{k_{\mathrm{iv1}}}{s}\right)\left(u_{\mathrm{fq}}^{*} - u_{\mathrm{fq}}\right) \\ i_{\mathrm{wq}}^{*} = \left(k_{\mathrm{pv2}} + \frac{k_{\mathrm{iv2}}}{s}\right)\left(u_{fd}^{*} - u_{\mathrm{fd}}\right) \end{cases} \tag{10}$$

where u_{fd}^* and u_{fq}^* are the *dq*-axis inverter voltage references, k_{pv1} and k_{iv1} are the parameters of the *q*-axis voltage control parameters, and k_{pv2} and k_{iv2} are the parameters of the *d*-axis voltage control parameters. Generally, the values of k_{pv1}, k_{iv1}, k_{pv2}, and k_{iv2} are chosen to ensure that the bandwidth of the voltage control loop is smaller than that of the current control loop for the sake of better system stability. In this paper, the inner current control bandwidth is designed as 300 Hz, which is smaller than $f_s/10$, where f_s is the sampling frequency of the inverter and the voltage control bandwidth is designed as one forth of the inner current control bandwidth.

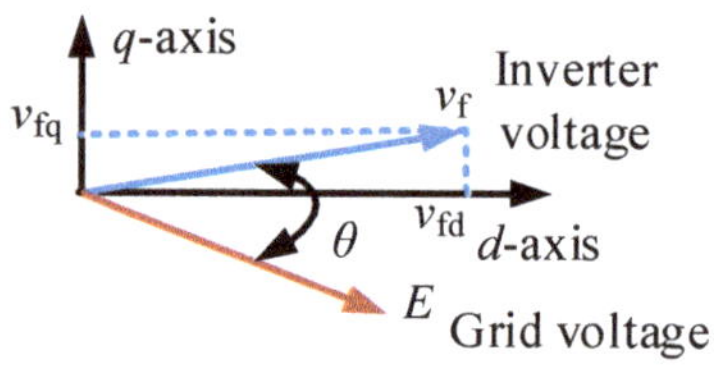

Figure 6. Voltage vector of the inverter and grid.

For the PLL-based frequency control, it is designed as

$$u_{fq}^* = k_f(f^* - f) \tag{11}$$

where f^* is the frequency reference, k_f is the proportional parameter for the frequency control. The frequency f obtained from the PLL is expressed as

$$\begin{cases} \omega_{pll} = k_{pp}u_{fq} + \frac{k_{ip}}{s}u_{fq} \\ f = \frac{\omega_{pll}}{2\pi} \end{cases} \tag{12}$$

where k_{pp} and k_{ip} are the parameters of the PLL and ω_{pll} is the frequency of the PLL. When the inverter frequency reference is larger than the measured one, the u_{fq}^* is larger than 0. With the combined action of the u_{fq} control loop and the PLL, the frequency and the phase angle will be increased. Thus, more active power can be exported. In contrast, when the inverter frequency reference is less than the measured one, the u_{fq}^* is smaller than 0. With the combined action of the u_{fq} control loop and the PLL, the frequency and the phase angle will be decreased. Thus, less active power will be exported.

For the inverter DC capacitor, the dynamic can be expressed as

$$C_{dc}\frac{du_{dc}}{dt} = i_{dcT} - i_{dc} \tag{13}$$

where C_{dc} is the DC capacitor value of the inverter, i_{dcT} is the DC current from the PV and BESS, and i_{dc} is the DC current of the inverter. As seen, the DC voltage is highly related to the DC current, which is related to the active power. Considering the connection between the frequency and the active power, the DC voltage control for the inverter is expressed as

$$f^* = \left(k_{pdc} + k_{idc}\frac{1}{s}\right)(u_{dc}^* - u_{dc}) + f_0 \tag{14}$$

where u_{dc}^* is the inverter DC voltage reference and k_{pdc} and k_{idc} are the parameters of the DC voltage control.

When a fault happens, the inverter switches from the grid-forming mode to the current-limiting mode automatically. The *dq*-axis current references of the inverter are based on the look-up table. The surplus power is absorbed by the BESS and when the DC voltage of the inverter is higher than the threshold value, the chopper inside the inverter will be triggered and the power will also be consumed partly by the resistor in the chopper. When the studied system is disconnected with the AC grid, the inverter is able to work on the islanded mode, as the control provides the frequency and AC voltage establishment capability.

Compared with the existing grid-forming control without the PLL, the proposed control strategy reserves the PLL. The main advantages of the proposed control are (1) when the offline renewable generation sources try to connect to the grid, the phase of the grid can be easily obtained for smooth connection and (2) when a fault happens, the current limiting control can be achieved smoothly through the automatically switching from the grid-forming control to the inner current control loop combined with PLL.

4. Simulation Result

In order to validate the effectiveness of the proposed control, a simulation model is established in PSCAD/EMTDC, as shown in Figure 7. The solar generation is modelled as aggregated model, which is rated at 300 MW. The battery energy storage (BES) is connected to the DC side of the solar generation through a buck/boost converter. The DC voltage of the battery is at 0.5 kV and the inverter DC voltage is at 1.25 kV. The rated capacity of the BES is 16.7 (kA·h). The other detailed parameters are presented in Table 1.

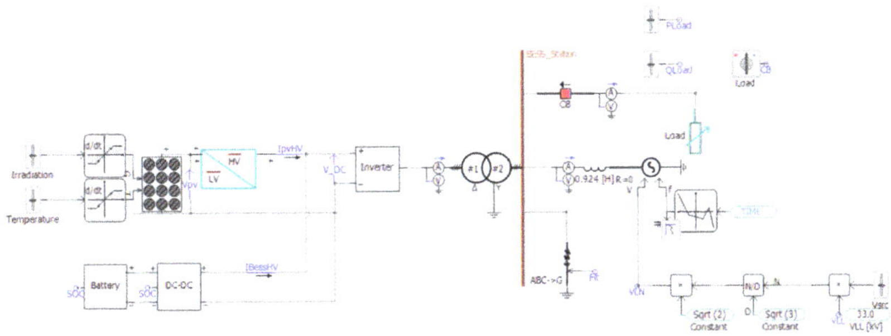

Figure 7. Simulation model in PSCAD/EMTDC of the studied system.

Table 1. Parameters of the system.

	Parameters	Values
BESS	DC voltage	0.5 kV
	Rated Capacity	16.7 (kA·h)
	Resistive Drop	0.001 pu
	Voltage at Exponential Point	1.03 pu
	Inductance and capacitor in DC/DC	0.2 μH 3000 mF 0.5 μH
Inverter	Rating	300 MVA
	DC voltage	1.25 kV
	AC voltage	0.69 kV
	DC capacitor	30,000 mF
	Transformer ratio	0.69/33 kV
	Transformer inductance and resistor	0.08, 0.002 pu
	Converter reactance and resistor	0.15 pu, 0.0015 pu
	Filter capacitor and resistor	0.10 pu, 0.001 pu

4.1. Performance Evaluation of the Grid-Forming Control during Power Variation and AC Voltage Variation

The performance of the grid-forming control is evaluated during the solar power generation variation and AC voltage variation. The timeline of the simulation is presented in Table 2.

Table 2. Sequence of simulation.

Time	Events
0–3 s	Solar power generation at 0.1 pu, BES operation off
3–3.5 s	Solar power generation increases from 0.1 pu to 1 pu
4.5–4.6 s	Solar power generation decreases from 1 pu to 0.2 pu
6–6.2 s	Inverter AC voltage decreases from 1 pu to 0.98 pu
8–8.2 s	Inverter AC voltage increases from 0.98 pu to 1.02 pu

Figure 8 shows the results of the system during such variations. As can be seen from Figure 8a, the solar power is initially generated at 30 MW before 3 s, and from 3 s to 3.5 s, the power is increased from 30 MW to 300 MW. During the active power increase, the inverter DC voltage and AC voltage can both be maintained at 1 pu, as shown in Figure 8c,d. The reactive power from the inverter sees a small increase during the power variation and is maintained at 8 MVar after 3.5 s. The magnitude of the AC current is increased from 0.035 MA to 0.35 MA, as presented in Figure 8e,f, which shows the DC current of the inverter, and the DC current is increased from 0.024 MA to 0.24 MA during 3 s to 3.5 s.

From 4.5 s to 4.6 s, the active power generation is decreased from 300 MW to 60 MW, as can be seen in Figure 8a. The DC current of the inverter is correspondingly decreased from 0.24 MA to 0.048 MA, as can be seen in Figure 8f. Figure 8c,d shows the DC voltage and AC voltage can be maintained at 1 pu after such an active power variation, although a small voltage oscillation is observed during the active power variation. Figure 8b shows the inverter reactive power. As can be seen, the reactive power is decreased from 8 MVAr to 0 MVAr from 4.5 s to 4.6 s. The magnitude of the inverter AC current decreases from 0.24 MA to 0.07 MA, as seen in Figure 8e.

From 6 s to 6.2 s, the inverter AC voltage reference decreases from 1 pu to 0.98 pu. As seen in Figure 8d, the measured AC voltage follows the reference and reaches 0.98 pu at 6.6 s. The measured AC voltage follows the reference decrease and shows the over-damped dynamic, indicating the strong stability of the voltage control. With the voltage variation, the inverter reactive power is decreased from 0 MVar to —100 MVar, as shown in Figure 8b. The inverter AC current magnitude is increased from 0.07 MA to 0.145 MA, as shown in Figure 8e. From 8 s to 8.2 s. the inverter AC voltage reference is increased from 0.98 pu to 1.02 pu, and the measured AC voltage follows the reference increase and is controlled at 1.02 pu after 8.5 s, as presented in Figure 8d. The inverter reactive power is increased from −100 MVAr to 105 MVAr, as shown in Figure 8b. The results are matched by the fact that when the AC voltage magnitude is increased, the inverter will output more reactive power. During the AC voltage variation, the AC current sees the initial decrease and then increase, as presented in Figure 8e. Such a performance is caused by the change in the absolute value of the reactive power. Figure 8a,c,f shows the performance of the inverter's active power, DC voltage, and DC current, respectively. As can be seen, all these variables experience little change during the inverter AC voltage variation.

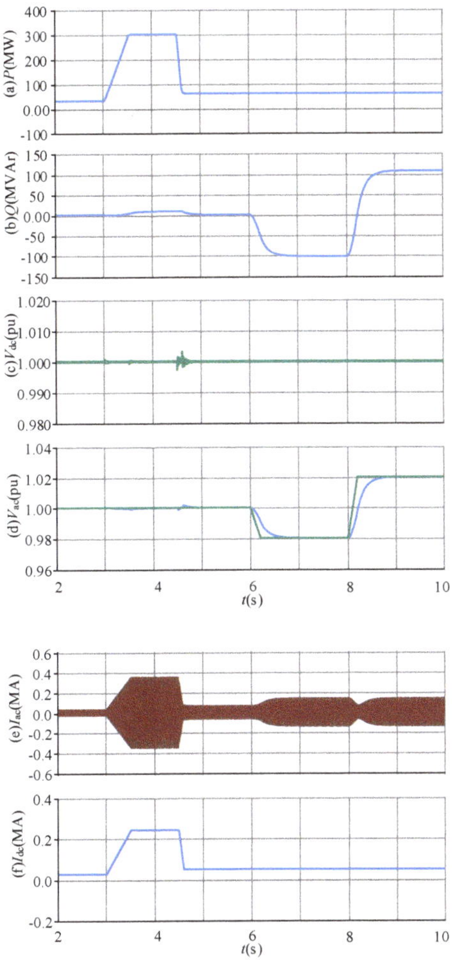

Figure 8. Simulation results during power variation and AC voltage variation: (**a**) inverter active power; (**b**) inverter reactive power; (**c**) inverter DC voltage; (**d**) inverter AC voltage (RMS); (**e**) inverter instantaneous AC current; (**f**) inverter DC current.

4.2. Performance Evaluation of BESS during the Grid Frequency Change

The performance of the BESS is evaluated during the grid frequency change and the corresponding results are presented in Figure 9. It is worth mentioning that the time sequence of this part is different from that of Section 4.1, so the simulation time in Figure 9 can be different from Figure 8.

From 2 s to 3 s, the grid frequency is decreased from 60 Hz to 59.6 Hz, as shown in Figure 9a. Due to the inertial and frequency support capability of the BESS, the inertia time constant of the system is 10 s and the primary frequency coefficient is 50 MW/Hz, and the active power from the BESS is increased from 0 MW to 20 MW, as presented in Figure 9b. Figure 9c,d shows the voltage of the BESS and the current of the BESS, respectively. As can be seen, the voltage is decreased from 556.8 V to 551 V and the current of the BESS is increased from 0 kA to 39 kA. Due to the discharging of the BESS, the SOC of the battery is decreased from the initial value of 94.98% at 2 s to 96.96% at 3 s, as shown in Figure 9e. With the power support from the BESS, the active power of the inverter is increased from the initial value of 150 MW to 170 MW, as shown in Figure 9f. It should be noted the initial power of 150 MW before 3 s is exported by solar generation. During the power increase from the BESS, the inverter DC voltage is increased from 1.25 kV to 1.272 kV, as shown in Figure 9g. Figure 9h shows the AC voltage of the inverter. As can be seen, the AC voltage is maintained at a constant value.

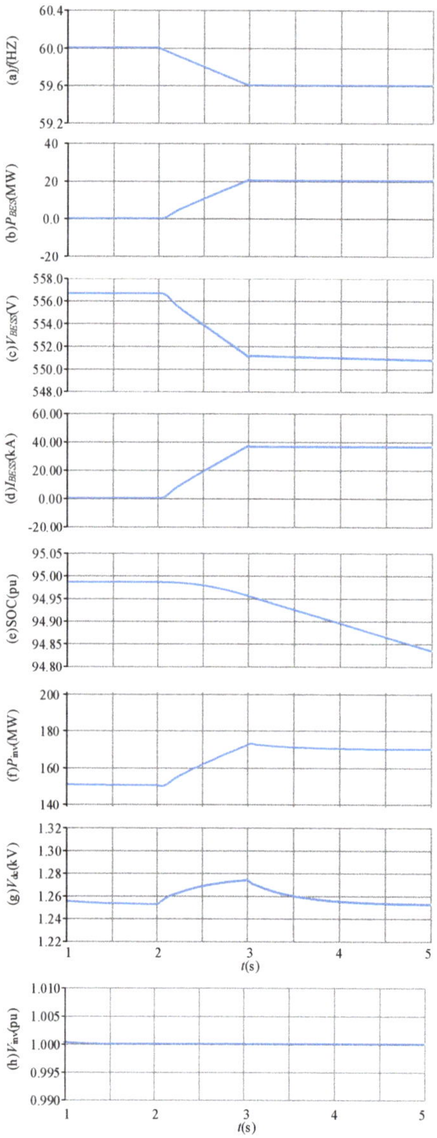

Figure 9. Simulation results of BESS during the grid frequency change: (**a**) grid frequency; (**b**) battery power; (**c**) battery voltage; (**d**) battery current; (**e**) state of charge; (**f**) inverter active power; (**g**) inverter DC voltage; (**h**) inverter AC voltage.

After 3 s, the grid frequency is constant at 59.6 Hz, as shown in Figure 9a. At this stage, the BESS keeps providing the active power for the primary frequency regulation and the exported active power is at 20 MW, as presented in Figure 9b–d, which demonstrates the voltage of the BESS and the current of the BESS, respectively. As can be seen, the voltage and the current are maintained at a constant value of 551 V and 39 kA. The SOC of the battery keeps decreasing and is at 94.84% at 5 s, as shown in Figure 9e–h, which shows that the inverter active power, DC voltage, and AC voltage can all operate stably after 3 s, at 170 MW, 1.25 kV, and 1 pu, respectively.

4.3. Performance Evaluation of System during the Grid Fault

The performance of the system during the grid fault is assessed and the results are presented in Figure 10. Before 3 s, the system operates at rated values with the active power at 300 MW, while the reactive power from the inverter is around 0 MVAr, as shown in Figure 10d,e. At 3 s, a three-phase fault happens and the voltage of the converter is decreased to 0.13 pu, as shown in Figure 10a,b, which shows the grid side instantaneous voltage, the magnitude of the voltage decreases from 27 kV to 3.4 kV, and the voltage is dominated by the fundamental frequency component. After the severe AC fault, the converter operates on the current-limiting control mode, and the converter AC current magnitude is controlled at 0.59 MA, as shown in Figure 10c. After 200 ms, the fault is cleared. As can be seen in Figure 10a,b, the voltage recovers to 1 pu. The inverter is no longer working on the current-limiting mode and recovers to transmit the active power. As seen in Figure 10d, the active power generation is quickly restored after the fault clearance. Figure 10e shows the reactive power recovery performance, which can also be restored to 0 MVAr, although its recovery speed is slower than the active power generation.

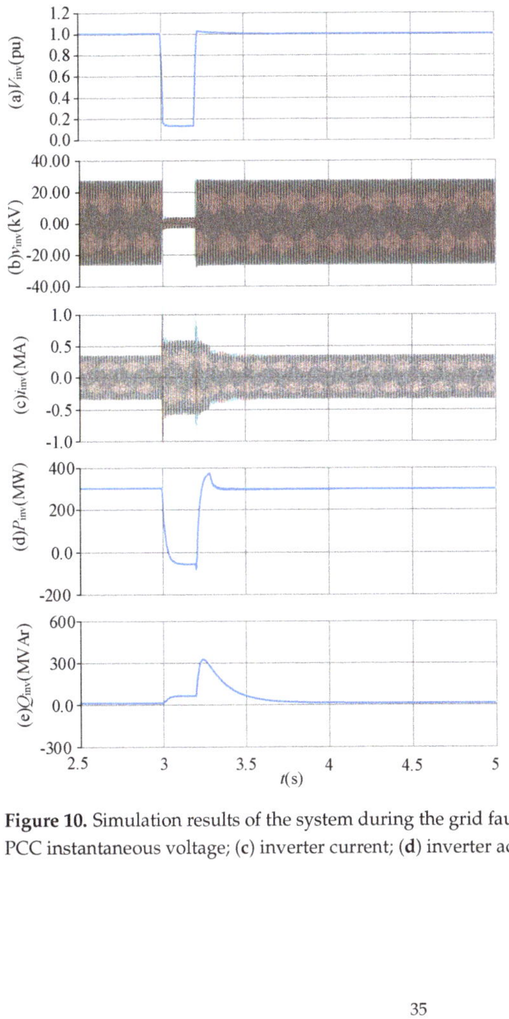

Figure 10. Simulation results of the system during the grid fault: (**a**) grid PCC voltage (RMS); (**b**) grid PCC instantaneous voltage; (**c**) inverter current; (**d**) inverter active power; (**e**) inverter reactive power.

4.4. Performance of System during the Transition from the Grid-Connected Mode to Islanded Mode

The performance of the system during the transition from the grid-connected mode to the islanded mode is evaluated and the results are presented in Figure 11. Before 3 s, the system works on the rated value, with the AC voltage at 1 pu, frequency at 60 Hz, DC voltage at 1.25 kV, active power at 300 MW, and reactive power at 3 MVar, as shown in Figure 11a–e. At 3 s, the inverter is disconnected from the grid. As can be seen in Figure 11a,b, the voltage and frequency experience some oscillation after the operation mode switch but can stabilize after 0.8 s. Figure 11e shows the reactive power output from the inverter. As seen, before and after the mode switch, the reactive power is decreased from 3.3 MVAr to 2 MVAr. Figure 11f shows the inverter current. As can be seen, the current is not disturbed by the mode switch.

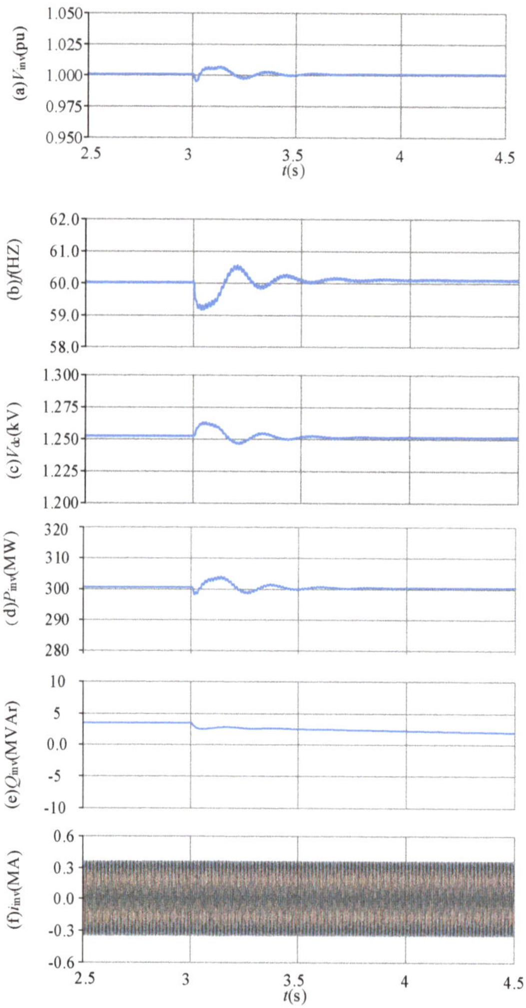

Figure 11. Simulation results during the transition from the grid-connected mode to islanded mode: (**a**) inverter AC voltage; (**b**) inverter frequency; (**c**) inverter DC voltage; (**d**) inverter active power; (**e**) inverter reactive power; (**f**) inverter AC instantaneous current.

4.5. Performance of System during Phase Jump and Grid Impedance Variation

The performance of the system during phase jump is evaluated and the results are presented in Figure 12. At 3 s, the phase of the grid experiences a sudden jump, with a relatively large variation value at 30°, as seen in Figure 12a. The active power and reactive power of the inverter drift away from the initial steady-state value after such a disturbance but can stabilize after 0.1 s, as shown in Figure 12b–e, which shows the DC voltage and AC voltage, respectively. As presented, the DC voltage decreases after the AC grid phase jump while the inverter AC voltage sees a small oscillation, but both can become stable again after the disturbance. Figure 12f shows the inverter AC instantaneous current. Due to the disturbance, the three-phase current sees some harmonic components in the first place but is able to become symmetrical quickly after 40 ms.

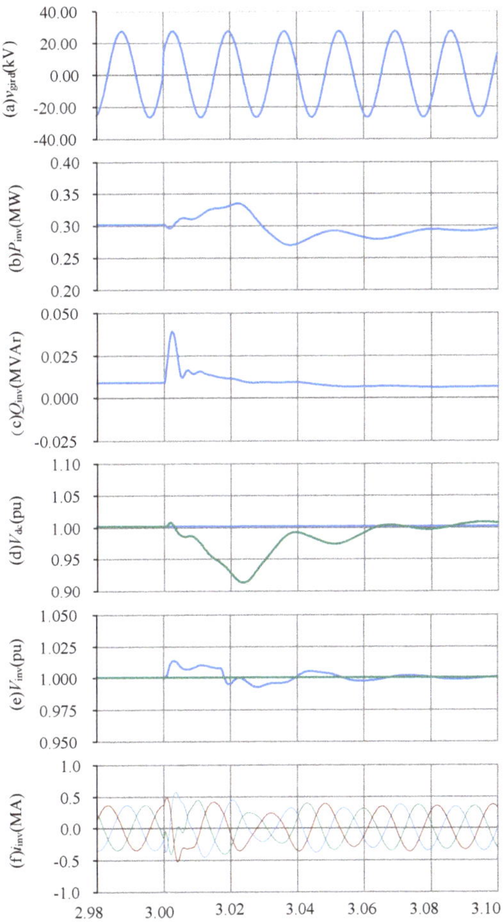

Figure 12. Simulation results during phase jump: (**a**) grid instantaneous AC voltage; (**b**) inverter active power; (**c**) inverter reactive power; (**d**) inverter DC voltage; (**e**) inverter AC voltage; (**f**) inverter AC instantaneous current.

The performance of the system during grid impedance variation is presented in Figure 13. At 3 s, the grid impedance is increased from 0.2 pu (SCR = 5) to 0.33 pu (SCR = 3), and at 4 s, the grid impedance is further increased to 0.5 pu (SCR = 2). As can be seen in Figure 13a,b, the proposed control has the capability to control the inverter active power and inverter AC voltage during different grid impedances.

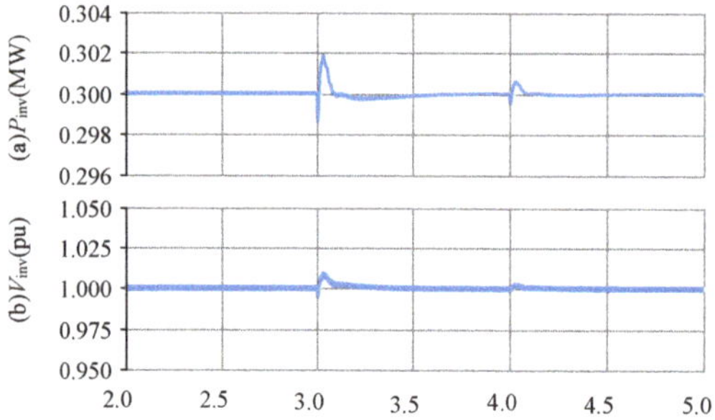

Figure 13. Simulation results during grid impedance ratio variation: (**a**) inverter active power; (**b**) inverter AC voltage.

5. Conclusions

This paper proposes a grid-forming control for solar generation systems with battery energy storage. The inverter control consists of the inner current control, voltage control, the PLL-based frequency control, and the DC voltage control. With the proposed control, the battery energy storage is able to provide inertial and primary frequency support during the grid frequency disturbance. Simulation models are established in PSCAD/EMTDC, and the results during the active power variation and AC variation, the grid frequency disturbance, grid fault, and mode switch validate the effectiveness of the proposed control.

Author Contributions: Methodology, Y.C. and M.W.; conceptualization and supervision, L.Y.; writing—original draft, S.L., Z.F. and W.T.; writing—review and editing W.L.; validation S.S. All authors have read and agreed to the published version of the manuscript.

Funding: This research is supported by the Science and Technology Project of State Grid Liaoning Electric Power Co., Ltd. (2023YF-90). The funder had the following involvement with the study: the study design and the decision to submit it for publication.

Data Availability Statement: The original contribution presented in this study is included in the article and further inquiries can be directed to the corresponding author.

Conflicts of Interest: Authors Yupeng Cai, Meng Wu, Shengyang Lv, Wei Li, and Songjie Shi were employed by the State Grid Liaoning Electric Power Co., Ltd. The remaining authors declare that the research was conducted in the absence of any commercial or financial relationships that could be construed as potential conflicts of interest.

References

1. Rocabert, J.; Luna, A.; Blaabjerg, F.; Rodríguez, P. Control of Power Converters in AC Microgrids. *IEEE Trans. Power Electron.* **2012**, *27*, 4734–4749. [CrossRef]
2. Pogaku, N.; Prodanovic, M.; Green, T.C. Modeling, Analysis and Testing of Autonomous Operation of an Inverter-Based Microgrid. *IEEE Trans. Power Electron.* **2007**, *22*, 613–625. [CrossRef]
3. Harnefors, L.; Wang, X.; Yepes, A.G.; Blaabjerg, F. Passivity-Based Stability Assessment of Grid-Connected VSCs—An Overview. *IEEE J. Emerg. Sel. Top. Power Electron.* **2016**, *4*, 116–125. [CrossRef]
4. Zhou, J.Z.; Ding, H.; Fan, S.; Zhang, Y.; Gole, A.M. Impact of Short-Circuit Ratio and Phase-Locked-Loop Parameters on the Small-Signal Behavior of a VSC-HVDC Converter. *IEEE Trans. Power Deliv.* **2014**, *29*, 2287–2296. [CrossRef]
5. Fan, L. Modeling Type-4 Wind in Weak Grids. *IEEE Trans. Sustain. Energy* **2019**, *10*, 853–864. [CrossRef]
6. Hu, J.; Huang, Y.; Wang, D.; Yuan, H.; Yuan, X. Modeling of Grid-Connected DFIG-Based Wind Turbines for DC-Link Voltage Stability Analysis. *IEEE Trans. Sustain. Energy* **2015**, *6*, 1325–1336. [CrossRef]
7. Lin, X.; Liu, Y.; Yu, J.; Yu, R.; Zhang, J.; Wen, H. Stability analysis of Three-phase Grid-Connected inverter under the weak grids with asymmetrical grid impedance by LTP theory in time domain. *Int. J. Electr. Power Energy Syst.* **2022**, *142*, 108244. [CrossRef]
8. Wen, B.; Boroyevich, D.; Burgos, R.; Mattavelli, P.; Shen, Z. Analysis of D-Q Small-Signal Impedance of Grid-Tied Inverters. *IEEE Trans. Power Electron.* **2016**, *31*, 675–687. [CrossRef]

9. Sun, J. Impedance-Based Stability Criterion for Grid-Connected Inverters. *IEEE Trans. Power Electron.* **2011**, *26*, 3075–3078. [CrossRef]
10. Zhang, X.; Xia, D.; Fu, Z.; Wang, G.; Xu, D. An Improved Feedforward Control Method Considering PLL Dynamics to Improve Weak Grid Stability of Grid-Connected Inverters. *IEEE Trans. Ind. Appl.* **2018**, *54*, 5143–5151. [CrossRef]
11. Givaki, K.; Chen, D.; Xu, L. Current Error Based Compensations for VSC Current Control in Weak Grids for Wind Farm Applications. *IEEE Trans. Sustain. Energy* **2019**, *10*, 26–35. [CrossRef]
12. Egea-Alvarez, A.; Fekriasl, S.; Hassan, F.; Gomis-Bellmunt, O. Advanced Vector Control for Voltage Source Converters Connected to Weak Grids. *IEEE Trans. Power Syst.* **2015**, *30*, 3072–3081. [CrossRef]
13. Zhang, L.; Harnefors, L.; Nee, H. Power-Synchronization Control of Grid-Connected Voltage-Source Converters. *IEEE Trans. Power Syst.* **2010**, *25*, 809–820. [CrossRef]
14. Zhang, L.; Harnefors, L.; Nee, H. Interconnection of Two Very Weak AC Systems by VSC-HVDC Links Using Power-Synchronization Control. *IEEE Trans. Power Syst.* **2011**, *26*, 344–355. [CrossRef]
15. Liu, J.; Miura, Y.; Ise, T. Comparison of Dynamic Characteristics Between Virtual Synchronous Generator and Droop Control in Inverter-Based Distributed Generators. *IEEE Trans. Power Electron.* **2016**, *31*, 3600–3611. [CrossRef]
16. Meng, X.; Liu, J.; Liu, Z. A Generalized Droop Control for Grid-Supporting Inverter Based on Comparison Between Traditional Droop Control and Virtual Synchronous Generator Control. *IEEE Trans. Power Electron.* **2019**, *34*, 5416–5438. [CrossRef]
17. Wu, W.; Chen, Y.; Zhou, L.; Luo, A.; Zhou, X.; He, Z.; Yang, L.; Xie, Z.; Liu, J.; Zhang, M. Sequence Impedance Modeling and Stability Comparative Analysis of Voltage-Controlled VSGs and Current-Controlled VSGs. *IEEE Trans. Ind. Electron.* **2019**, *66*, 6460–6472. [CrossRef]
18. Li, M.; Wang, Y.; Hu, W.; Shu, S.; Yu, P.; Zhang, Z.; Blaabjerg, F. Unified Modeling and Analysis of Dynamic Power Coupling for Grid-Forming Converters. *IEEE Trans. Power Electron.* **2022**, *37*, 2321–2337. [CrossRef]
19. Mallemaci, V.; Mandrile, F.; Rubino, S.; Mazza, A.; Carpaneto, E.; Bojoi, R. A comprehensive comparison of Virtual Synchronous Generators with focus on virtual inertia and frequency regulation. *Electr. Power Syst. Res.* **2021**, *201*, 107516. [CrossRef]
20. Barać, B.; Krpan, M.; Capuder, T.; Kuzle, I. Modeling and Initialization of a Virtual Synchronous Machine for Power System Fundamental Frequency Simulations. *IEEE Access* **2021**, *9*, 160116–160134. [CrossRef]

energies

MDPI

Article

Optimization of PID Controllers Using Groupers and Moray Eels Optimization with Dual-Stream Multi-Dependency Graph Neural Networks for Enhanced Dynamic Performance

Vaishali H. Kamble [1], Manisha Dale [2], R. B. Dhumale [3] and Aziz Nanthaamornphong [4,*]

1 Department of Electronics and Communication Engineering, DES Pune University, Pune 411004, India; vaishaliraichurkar@gmail.com
2 Department of Electronics and Telecommunication, MES Wadia College of Engineering, Pune 411004, India; mpdale@mescoepune.org
3 Department of Electronics and Telecommunication Engineering, AISSMS Institute of Information Technology, Pune 411001, India; rbd.scoe@gmail.com
4 College of Computing, Prince of Songkla University, Phuket 83120, Thailand
* Correspondence: aziz.n@phuket.psu.ac.th

Abstract: Traditional proportional–integral–derivative (PID) controllers are often utilized in industrial control applications due to their simplicity and ease of implementation. This study presents a novel control strategy that integrates the Groupers and Moray Eels Optimization (GMEO) algorithm with a Dual-Stream Multi-Dependency Graph Neural Network (DMGNN) to optimize PID controller parameters. The approach addresses key challenges such as system nonlinearity, dynamic adaptation to fluctuating conditions, and maintaining robust performance. In the proposed framework, the GMEO technique is employed to optimize the PID gain values, while the DMGNN model forecasts system behavior and enables localized adjustments to the PID parameters based on feedback. This dynamic tuning mechanism enables the controller to adapt effectively to changes in input voltage and load variations, thereby enhancing system accuracy, responsiveness, and overall performance. The proposed strategy is assessed and contrasted with existing strategies on the MATLAB platform. The proposed system achieves a significantly reduced settling time of 100 ms, ensuring rapid response and stability under varying load conditions. Additionally, it minimizes overshoot to 1.5% and reduces the steady-state error to just 0.005 V, demonstrating superior accuracy and efficiency compared to existing methods. These improvements demonstrate the system's ability to deliver optimal performance while effectively adapting to dynamic environments, showcasing its superiority over existing techniques.

Keywords: buck-boost converters; proportional integral derivative; Schottky diode; error signal; control signal; steady-state error; tuning methods

1. Introduction

The simplicity, convenience of use, and typically satisfactory performance of proportional–integral–derivative (PID) controllers make them popular in industrial applications [1]. These controllers are often employed in power electronic systems, like DC–DC converters and inverters, for voltage, current, and power regulation [2]. DC–DC converters are especially important in modern energy systems due to their role in efficiently managing and converting power from renewable sources like solar panels and batteries [3]. They enable voltage regulation and energy transfer across different subsystems, improving overall energy efficiency and system reliability [4]. These converters are integral in electric vehicles, smart grids, and portable electronic devices, where consistent and adaptive power delivery is critical. Their capacity to provide controlled voltage levels makes them essential for integrating intermittent and variable energy sources into stable usable outputs [5]. The traditional PID tuning methods like Ziegler–Nichols and Cohen–Koon often yield suboptimal results in these nonlinear systems, as they are designed for linear or time-invariant conditions [6]. Consequently, conventional PID controllers, which are typically tuned for steady-state performance [7], may fail to provide an adequate transient response or stability in applications where load or input voltage fluctuates [8]. Moreover, in the presence of high-frequency switching noise, which is characteristic of buck-boost converters, improper PID tuning can lead to performance degradation such as overshooting or instability [9]. The control process can become more complicated when the system is subjected to fluctuating load conditions, making it crucial to adjust PID parameters to maintain optimal performance [10]. Recent advancements have focused on adaptive PID control strategies, which dynamically adjust the controller parameters to handle the system's nonlinearities and improve its robustness [11]. One approach is using optimization techniques like particle swarm optimization or genetic algorithms, which fine-tune PID parameters to achieve better performance in nonlinear and time-varying environments [12]. An additional promising strategy involves incorporating artificial intelligence or machine learning models to dynamically adjust PID gains, enhancing the controller's adaptability [13]. Despite these advancements, traditional PID control still faces limitations in achieving optimal performance under highly dynamic and nonlinear conditions [14].

In buck-boost converters, the PID controller must maintain a delicate balance between stability, transient response, and steady-state accuracy, which is often challenging due to the varying dynamics of the system [15]. A key challenge lies in selecting PID gains that remain effective across a variety of operational circumstances, including different load levels and input voltages [16]. In order to improve the performance of PID controllers in buck-boost converters, current research is still investigating hybrid and adaptive control techniques [17]. The importance of continuous adjustment and fine-tuning of PID parameters becomes evident as the converter's dynamic response must be matched with external changes, requiring constant optimization [18]. High-frequency noise from the switching operation of the converter can further complicate the controller's performance, leading to higher overshoot and instability unless properly addressed [19]. Therefore, more sophisticated approaches that integrate global optimization and adaptive feedback mechanisms are crucial for improving the overall stability and efficiency of PID-controlled buck-boost converters under dynamic conditions [20].

In the literature, various research works are available based on PID controller optimization, adaptive control strategies, and performance enhancement in buck-boost converters using various techniques and aspects. A few of these works are reviewed as follows.

1.1. Optimization Techniques

S Sangeetha et al. [21] suggested a hybrid technique for fractional-order proportional integral derivative (FOPID) controller performance analysis of buck converters. The Capuchin Search Algorithm (CapSA) and the Golden Jackal Optimization (GJO) were integrated into this hybrid technique. The Capuchin Search Algorithm was used to improve the Golden Jackal Optimization's update behavior, resulting in the enhanced GJO (IGJO) approach. Because power converters were nonlinear, they were difficult to regulate, and there was a constant quest for efficient and effective controllers. Recently, it has been demonstrated that fractional-order controllers are more efficient in power electronic systems. The best design for a fractional-order PID controller for the buck converter was found using the IGJO approach.

P Warrier et al. [22] presented a complex-order PI controller for DC–DC buck and boost converter control that combines a complex-order integrator. Four parameters in the intricate PID controller require adjustment. The Metaheuristic Cohort Intelligence method was used to optimize the design of the complex-order PI controller. The outcomes were contrasted with those of a PID controller of fractional order. The results showed that the complex PI controller provided a better response than the FOPID controller and was more robust to parameter changes.

N F Nanyan et al. [23] presented an improved Sine Cosine Algorithm (ISCA) for the optimization of a DC–DC buck converter using a PID controller. Through two distinct improvements, the limitations of the traditional Sine Cosine Algorithm (SCA) were addressed, resulting in a synergistic usage of nonlinear equations in the instrumental mechanism to revise the average location. To address the problem of local optima, the initial revision included an instrumental function to update the average location. The second change applied a nonlinear equation to the algorithm's reducing position-updating mechanism, coordinating the traditional SCA's disproportional exploration and exploitation phases.

L.K. Fong et al. [24] proposed the Archimedes Optimization Algorithm (AOA) as a metaheuristic method for optimizing a PID controller in a closed-loop DC–DC buck converter. The primary duty of the converter was to control output voltage, ensuring stability in the face of input voltage variations and load variations. The performance of the converter heavily depends on the PID controller's gain settings, which need to be optimized for robust operation. AOA helps identify the optimal integral, proportional, and derivative gains, improving the controller's ability to minimize voltage errors, reduce overshoot, and enhance response time across varying load conditions. However, AOA may require significant computational resources for complex systems.

1.2. Controller Structure

S M Ghamari et al. [25] designed a Lyapunov-based model reference PID controller for a DC/DC buck converter using their approach. For more dependable functioning, the parameters must be returned, and the PID approach is unsuitable for real-world applications due to a variety of disruptions. To solve this, the PID approach used an adaptive mechanism based on the Lyapunov definition, which improved the stability of the system and resilience to a range of shocks. The system was also handled as a "black-box", which removes the requirement for precise mathematical modeling and eases installation and computing load. As a contemporary adaptive algorithm, the Lyapunov notion may provide optimum solutions more quickly while maintaining dependable stability and accuracy.

P. Sharma et al. [26] suggested an optimized FOPID controller, designed to generate optimal switching signals for the converter based on input voltage. The controller's performance was evaluated for both buck and boost modes of operation. It was verified that zero voltage switching (ZVS) and zero current switching (ZCS) are achieved under boost mode, enhancing efficiency and reducing switching losses. However, the implementation of the FOPID controller can be complex and requires high computational resources, especially in systems with varying load conditions.

D M Acevedo et al. [27] have presented an adaptive speed control strategy for DC motors using a DC/DC converter (in buck and boost modes) based on the inverse optimal control (IOC) method. Their method ensures stability and optimal performance of nonlinear systems through Lyapunov theory, minimizing a specified cost function. An integral action enhances performance, ensuring asymptotic stability without affecting convergence properties. The strategy was implemented on buck and boost converter/DC motor systems. However, the IOC approach can be computationally intensive and may not be ideal for systems with rapidly changing dynamics. Table 1 displays the summary of the research work.

Table 1. The summary of research work.

Authors	Methodology	Advantages	Limitations
S Sangeetha et al. [21]	Improved Golden Jackal Optimization (IGJO) for PID control	Efficient in power electronic systems, improved fractional-order control	Nonlinear nature of power converters, complexity in controller tuning
P Warrier et al. [22]	Cohort Intelligence Algorithm for PID control	More robust to parameter variations	Higher parameter complexity, challenging for practical implementation
N F Nanyan et al. [23]	Improved Sine Cosine Algorithm (ISCA) for PID control	Better local optima handling, enhanced PID tuning	Limited to specific types of systems, less efficient for highly nonlinear systems
L.K. Fong et al. [24]	Genetic Algorithm-based PID control	Effective for optimizing PID parameters in complex systems	Requires significant computational effort and time
S M Ghamari et al. [25]	Lyapunov-based model reference PID control	Improved stability, robustness under disturbances	May not work efficiently for rapid dynamic changes
P. Sharma et al. [26]	Firefly Algorithm for PID control	Global optimization, effective for nonlinear systems.	Require fine-tuning of parameters
D M Acevedo et al. [27]	Ant Colony Optimization for PID control	Effective in finding optimal PID gains, suitable for complex systems	High computational cost, sensitivity to initialization parameters

The generic review of recent research highlights various optimization techniques designed to enhance the control systems' performance, particularly in buck-boost converters, by optimizing PID controllers for better stability, efficiency, and dynamic adaptation under varying load conditions. The existing techniques include Improved Golden Jackal Optimization (IGJO) optimized PID, Improved Sine Cosine Algorithm (ISCA) optimized PID, Cohort Intelligence Algorithm (CTA) optimized PID, and Lyapunov-based adaptive PID (L-Based Adaptive PID). The IGJO-Optimized PID suffers from slow convergence, high overshoot, higher steady-state error, and sensitivity to initial conditions, making it less suitable for precision-critical applications. It also struggles to maintain stability in systems with significant variations in load or input, reducing its reliability in dynamic environments. The ISCA-Optimized PID faces challenges with response speed and precision, and its optimization process can be computationally expensive, limiting scalability for larger or more complex systems. Additionally, it may not handle nonlinearities effectively. The CTA-Optimized PID has limited adaptation to highly dynamic conditions, resulting in slower adjustments in unpredictable systems, and reduced robustness under external disturbances, leading to performance degradation. The L-Based Adaptive PID relies on Lyapunov functions, limiting its application to specific system types. It also exhibits slower adaptation in dynamic environments, higher computational complexity, and reduced effectiveness in nonlinear systems, making it less practical for fast-changing or highly nonlinear applications. Very few approach-based studies are offered in the literature to deal with this problem; these issues and disadvantages served as the impetus for this study work.

The proposed GMEO-DMGNN method was chosen due to its unique ability to address the complex optimization challenges in PID controllers for buck-boost converters. This combination overcomes existing drawbacks by offering better adaptability, faster convergence, and enhanced stability compared to conventional methods. Advantages include improved stability, reduced overshoot, minimized steady-state error, and faster settling times under varying loads. The novelty lies in integrating the GMEO with the DMGNN model for fine-tuning PID parameters, ensuring superior performance. The method works by leveraging GMEO for global search and DMGNN for localized fine-tuning of PID parameters, enhancing efficiency in dynamic systems. Compared to existing techniques, the GMEO-DMGNN method ensures better adaptability, responsiveness, and optimal performance in buck-boost converters under changing conditions, making it a more robust and efficient solution.

The following are the paper's primary contributions:

- This work uses GMEO and DMGNN to create an adaptive tuning mechanism for PID-controlled buck-boost converters, evaluated across dynamic loads, disturbances, and diverse signal profile;
- The proposed system utilizes a standard PID controller designed specifically for buck-boost converters, with enhancements that focus on optimizing its parameters for improved performance. The proposed method effectively addresses the nonlinear behavior and feedback noise commonly present in buck-boost converters;
- Through the integration of GMEO and DMGNN, the control system achieves a notable reduction in settling time of 100 ms, delivering an efficient response even under rapidly shifting operational conditions;
- The proposed method effectively stabilizes the nonlinear voltage gain characteristics of the buck-boost converter, which vary with the duty cycle, ensuring consistent output regulation across a wide range of input conditions and operating modes.

The remainder of the document is structured as follows: Section 2 clarifies the configuration for Optimizing PID Controllers in buck-boost Converters. Section 3 discusses the Simulation Configurations and Setup. Section 4 discusses the proposed GMEO-DMGNN. The results and discussion are clarified in Section 5, and Section 6 contains the conclusions.

2. Configuration for Optimizing PID Controllers in Buck-Boost Converters

Figure 1 illustrates the block diagram for optimizing PID controllers in buck-boost converters. It shows that the reference voltage is compared with the output voltage through a feedback loop to regulate the system. Subtracting from V_{ref} creates the error signal, which is then sent to the PID controller. Based on the error signal, the PID controller modifies the control signal $u(t)$, which is then used to regulate the buck-boost converter's output voltage V_{out}. The PID controller's performance is improved by optimizing the PID parameters through the integration of the proposed GMEO-DMGNN technique. The error signal and the control signal are dynamically adjusted through the GMEO-DMGNN method, which fine-tunes the PID parameters for optimal performance under varying load conditions. The system's stability and efficiency are increased by the integration of GMEO-DMGNN, which makes sure that the output voltage is kept at the intended set point while adjusting to dynamic changes. The GMEO-DMGNN method works by continually fine-tuning the PID parameters, allowing the system to adapt to these changing conditions. As a result, the system is able to maintain a steady output voltage (V_{out}) that is as close as possible to the reference voltage (V_{ref}), while also improving the entire stability, efficiency, and performance of the converter. Through this method, the proposed system not only ensures precise voltage regulation but also reduces overshoot, improves settling time, and enhances system robustness under various operating scenarios.

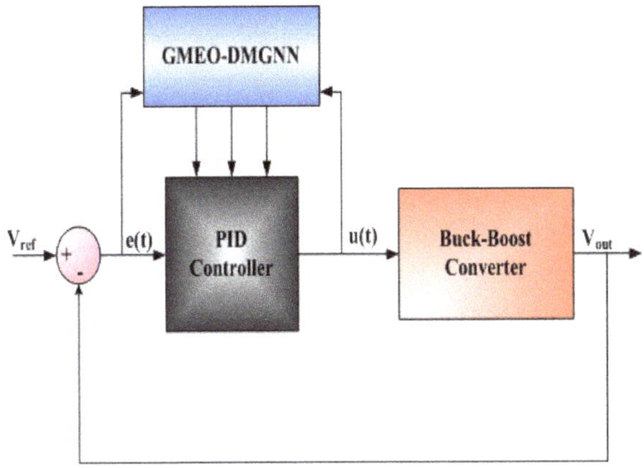

Figure 1. Block diagram of optimizing PID controllers in buck-boost converters.

2.1. Modeling of the Buck-Boost Converter

A buck-boost converter is a type of DC–DC converter that integrates the fundamental principles of both bucks and boost converters in one circuit as shown in Figure 2. The output voltage of this converter might be more or lower than the input voltage. The modeling used in this study is adapted from [28].

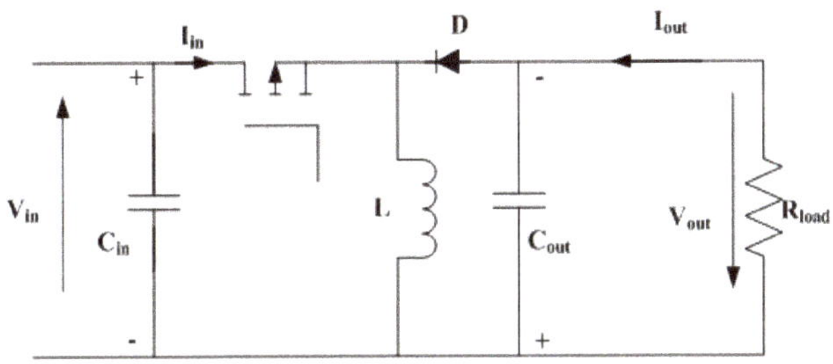

Figure 2. Buck-boost converter.

The energy balance method may be used to find the optimal value for the load resistor R_L, assuming the converter runs in boost mode with optimal efficiency.

$$R_L = \frac{V_{\max}^2}{P_{\max}} \quad (1)$$

The boost converter's voltage gain equation while it operates in continuous conduction mode (CCM) may be used to determine the duty cycle, while the energy balance concept is used to calculate the load resistor.

$$D = \frac{V_{\max}}{V_{\max} + V_{P\max}} \quad (2)$$

where $V_{\max}$, the input voltage, is given, and the duty cycle D is then calculated. The output voltage is defined in Equation (3)

$$V_{out} = \frac{V_{in} \cdot D}{1 - D} \quad (3)$$

where V_{out} is the output voltage. Once the duty cycle has been established, the inductor is made to restrict the input current ripple ΔI_{L_Pmax}.

$$L = \frac{V_{Pmax} \times D}{\Delta I_{L\_Pmax} \times fs} \quad (4)$$

where the current ripple ΔI_{L_Pmax} and the switching frequency fs are established, the input inductor's inductance is then calculated. The subsequent equation may be used to determine the input capacitor, C_{in}, as follows:

$$C_{in} = \frac{\Delta I_{L\_Pmax}}{\Delta V_{in\_Pmax} \times fs} \quad (5)$$

ΔV_{in_Pmax} is the minimum input voltage. The output capacitor is given in Equation (5)

$$C_{out} = \frac{I_{max} \times D}{\Delta V_{max} \times fs} \quad (6)$$

ΔV_{max} is the maximum output voltage. This guarantees a thorough and accurate depiction of the converter's operation at different load levels. The transfer function from the control input (duty ratio D) to the output is expressed as follows:

$$G_s = \frac{V_{OS}}{D_S} = \frac{K'}{(L_s + R) + (C_S + 1)} \quad (7)$$

where R represents the load resistance, L is inductance, and C represents the capacitance of the converter.

2.2. PID Controller Design

The standard PID controller is favored for its simplicity and reliability, making it suitable for a wide range of industrial applications. Compared to more complex controllers such as sigmoid-PID, FOPID, and Bandwidth-Enhanced Linear Bias Integral Controller (BELBIC), the standard PID controller offers easier implementation, reduced computational burden, and greater compatibility with existing systems. While advanced controllers may provide improved performance in specific scenarios, their complexity and tuning challenges often limit their practical use.

A feedback control loop is used by a PID controller to make clear linkages between the system variables, minimize the impact of disturbances, and guide the system toward the desired state [29]. The error at time t, represented by e_t, is sent into the controller. The discrepancy between the measured and reference values is known as this error. The action applied to the system or plant is represented by the PID controller's output, a_t. The integral gain K_i multiplied by the error's integral, the proportional gain K_p multiplied by the error magnitude, and the derivative gain K_d multiplied by the error's derivative are the three terms that make up the control signal or actuation. Some of these terms may be set to zero.

$$u(t) = K_p e_t + K_i \int_0^t e_t dt + K_d \frac{de_t}{dt} \quad (8)$$

The control signal $u(t)$, which is the total of the P, I, and D terms, is returned by the PID controller. In this case, K_p denotes proportional gain, K_d stands for derivative gain, and K_i denotes integral gain.

The PID controller's transfer function is described in Equation (9), as follows:

$$C(s) = K_p + \frac{K_i}{s} + K_d s \quad (9)$$

$C(s)$ is the transfer function of the PID controller in the Laplace domain.

The frequency response of the PID controller is given in Equation (10)

$$C(j\omega) = K_p + \frac{K_i}{j\omega} + K_d(j\omega) = K_p - \frac{jK_i}{j\omega} + jK_d(j\omega) \tag{10}$$

$C(j\omega)$ is the PID controller's frequency response. ω is the input signal's angular frequency in radians per second.

Shortly, the controller can be shown with its gain and phase values, as follows:

$$C(j\omega) = |C(j\omega)|e^{j<C(j\omega)} \tag{11}$$

$e^{j<C(j\omega)}$ iseuler's formula representation of a phase shift.

The gain value of the PID controller's frequency response is given in Equation (12)

$$|C(j\omega)| = \sqrt{K_p^2 + (-\frac{K_i}{\omega} + K_d\omega)^2} \tag{12}$$

The phase value of the controller's frequency response is given in Equation (13)

$$< C(j\omega) = \arctan\left(\frac{-\frac{K_i}{\omega} + K_d\omega}{K_p}\right) \tag{13}$$

where arctan is the inverse tangent function that calculates the phase shift. Thus, the frequency response of the PID controller is obtained. The initial values of the PID parameter are given in Table 2.

Table 2. The initial values of the PID parameter.

PID Parameter	Range
K_p	0.8 to 1.5
K_i	1000 to 3000
K_d	0.0001 to 0.001

3. Simulation Configurations and Setup

The buck-boost converter is developed with suitable properties before incorporating optimization techniques into a PID controller. The converter can handle the desired input and output voltages, loads, and switching frequencies, and certain characteristics and factor values are given in Table 3. An essential part of a buck-boost converter, the inductor (L), is responsible for both storing and transferring energy.

Table 3. The system specifications.

Parameters	Values
Input Voltage Range (V_{in})	5 V to 15 V
Output Voltage (V_{out})	12 V
Maximum Output Current (I_{out})	2 A
Switching Frequency (f_s)	100 Khz
Duty Cycle (D)	0.2 to 0.8

The value of the inductor has significant effects on the converter's performance and efficiency. The Equation (14) is used to obtain the inductance value.

$$L = \frac{\lfloor V_{in}(1 - D) \rfloor}{f_s \Delta_L} \tag{14}$$

Here, I_L indicates the highest output current that can pass through the inductor, usually set between 20% and 40% of the total allowable ripple current. The assumption that (for 40% ripple) $\Delta I_L = 0.4 I_{Out} L$ is given Equation (15).

$$L = \frac{\lfloor 15\text{V}(1 - 0.6) \rfloor}{100 \text{ Khz } 0.4 \times 2\text{A}} \approx 150 \ \mu\text{H} \tag{15}$$

The selected value of L is 150 µH. An output capacitor is required to lessen voltage ripple at the buck-boost converter's output. This helps stabilize the output by reducing voltage swings and removing high-frequency switching noise. Using the Equation (16), the capacitance value is determined based on the permitted ripple voltage, inductor ripple current, and switching frequency.

$$C_{out} = \frac{\Delta I_L}{8 \cdot f_s \Delta_{out}} \approx 1.25 \ \mu\text{f} \tag{16}$$

The output voltage remains stable with the correct capacitance, ensuring reliable converter performance. In a low-side switching setup, a MOSFET (e.g., IRF540N, 100 V, 33 A) handles the maximum input voltage and current. A Schottky diode (1N5822) is selected for its fast recovery time and high current handling with minimal forward voltage drop. The LM5118 IC controls Pulse Width Modulation (PWM) switching and feedback. Resistors set the output and reference voltages and feed back to the control loop. After initial tuning, the GMEO algorithm combined with the DMGNN model adaptively refines the PID constants to enhance control efficiency and system performance.

$$G_s = \frac{V_{OS}}{D_s} = \frac{V_{in}(1 - D)}{(L \cdot C \cdot S^2) + (R \cdot C \cdot s) + 1} \tag{17}$$

The initial values for K_p, K_i, and K_d are obtained using the Ziegler–Nichols method and further fine-tuned through simulation to finalize the PID controller design. To ensure optimal performance across varying operating conditions, machine learning techniques using the GMEO-DMGNN framework are employed to dynamically adjust these parameters. The proposed controller continuously updates the gains based on system data, such as fluctuations in input voltage or load changes, effectively minimizing output voltage errors.

4. Proposed GMEO-DMGNN Method for Enhancing PID Control in Buck-Boost Converters

This section outlines the integration of a Dual-Stream Multi-Dependency Graph Neural Network (DMGNN) with the Groupers and Moray Eels (GMEO) for optimizing PID controller parameters. GMEO is used to optimize PID gains (K_p, K_i and K_d), while the Dual-Stream Multi-Dependency Graph Neural Network (DMGNN) predicts and locally adjusts these parameters to enhance performance. GMEO explores large solution spaces, maintains diversity to avoid local optima, and handles nonlinear systems to improve PID controller performance by enhancing stability, reducing overshoot, and optimizing response time. DMGNN further refines optimization by capturing complex dependencies and learning both global and local patterns, which accelerates convergence and improves performance in dynamic systems like buck-boost converters. The combination of GMEO's global search with DMGNN's local adjustments optimizes PID parameters more efficiently, improving system stability, response time, and adaptability while ensuring faster convergence in complex dynamic systems. Figure 3 depicts the flowchart of the GMEO-DMGNN approach.

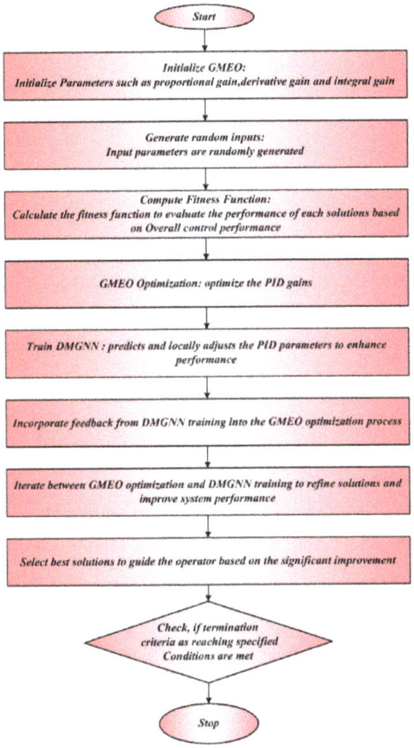

Figure 3. Flowchart of GMEO-DMGNN.

4.1. Optimization Using Groupers and Moray Eels (GMEO)

In this section, the GMEO is described [30] and utilized to optimize the controller parameters of PID gains, such as, K_d, K_p, and K_i. The GMEO algorithm offers a robust global search mechanism, efficiently optimizing PID parameters to enhance performance in nonlinear dynamic systems, ensuring improved stability, faster response, and better adaptability in buck-boost converters. In buck-boost converters, it increases overall system stability and performance by improving the PID controller's capacity to adjust to changing system circumstances, lowering overshoot, settling time, and steady-state error. GMEO was chosen for its ability to effectively handle the complexities of nonlinear dynamic systems by offering a balanced global search approach, optimizing PID parameters for improved control and performance in buck-boost converters.

Step 1: Initialization

Set the input variables to first values. In this instance, the input variables are the PID parameters, which are specified as K_d, K_p, and K_i.

Step 2: Random Generation

In matrix form, the input variables were generated at random.

$$q = \begin{bmatrix} k_{1,1} & k_{1,2} & \dots & k_{1,m} \\ k_{2,1} & k_{2,2} & \dots & k_{2,m} \\ k_{3,1} & k_{3,2} & \dots & k_{3,m} \end{bmatrix} \tag{18}$$

where q indicates the random generation, k indicates the system parameters, and m indicates the count of decision variables.

Step 3: Fitness Function

The fitness was evaluated, which was described by

$$F = \min(J) \tag{19}$$

$$J = \int_0^T t|e_t|dt \tag{20}$$

where J refers to an Integral of Time-Weighted Absolute Error (ITAE), t specifies the time variable, and e_t specifies the error signal at time t.

Step 4: Primary Search (PS) Phase

GMEO agents explore the search space for optimal PID controller parameters (K_d, K_p, and K_i), mimicking the zigzag swimming pattern of groupers hunting prey. This random exploration ensures thorough coverage of the solution space, aiming to find the optimal controller gains that minimize performance errors like transient response, steady-state error, and overshoot.

$$X_{ij}^{initial} = low_j + rand \cdot (upper_j - low_j), i = 1,2,3 \ldots N, j = 1,2,3, \ldots D \tag{21}$$

Here, $X_{ij}^{initial}$ specifies the first location of ith search agent of jth dimension, $upper_j$ and low_j specify the search space's upper and lower bounds, D specifies the overall count of dimensions, N specifies the number of search agents, and rand specifies a random vector that follows a uniform distribution, with values ranging from 0 to 1.

Step 5: Pair Association (PA) Phase

In this phase, the best-performing agents (groupers) collaborate with other high-quality agents (moray eels) to improve search efficiency. This cooperative interaction enhances the exploration of promising regions in the solution space. By leveraging the strengths of both agents, the search process becomes more targeted, accelerating convergence toward the optimal PID parameters. The agents dynamically adjust their positions based on the most promising solutions, ensuring a balance between global exploration and local refinement for improved accuracy in optimizing the PID controller.

Step 6: Encircling or Extended Search (ES) Phase

Agents refine their search by adaptively adjusting their positions toward promising regions. This phase enhances local exploration, allowing agents to dynamically focus on areas with higher potential for optimal PID parameters. The cooperative movement mimics the coordinated behavior of groupers and moray eels, ensuring a balance between exploitation and exploration. This adaptive search approach improves the likelihood of finding the global optimum and helps avoid premature convergence.

$$c_{mj} = X_{gmj} + \frac{L}{dis}(X_{Emj} - X_{gmj}) \tag{22}$$

Here, c_{mj} denotes the coordinates of the mth prey in each dimension, X_{gmj} specifies the location of a grouper, X_{Emj} specifies the location of an eel, L specifies the separation between the prey and the grouper, and dis specifies the distance between the grouper and the eel.

Step 7: Attacking and Catching Phase

Agents converge on the best solution by intensifying the search around the optimal PID gains. This phase improves convergence accuracy by gradually reducing the search radius, ensuring precise identification of the optimal controller parameters. The shrinking mechanism enables a finer search around the most promising solution, refining the PID gains for better system performance. This stage also helps to reduce the steady-state error and enhances system stability by continuously updating the solution based on the best-performing agents.

$$R_{i+1} = (1 - \mu) * R_i \tag{23}$$

where $i = 1, 2, 3, \ldots AB - 1$ and μ specifies the shrinking ratio and R refers to the radius.

Step 8: Termination Criteria

The procedure ends if the answer is ideal; if not, it goes back to step 3 for fitness assessments and keeps processing the next steps until the best answer is discovered. Thus, GMEO effectively optimized the controller parameters of PID gains. The flowchart of GMEO is illustrated in Figure 4.

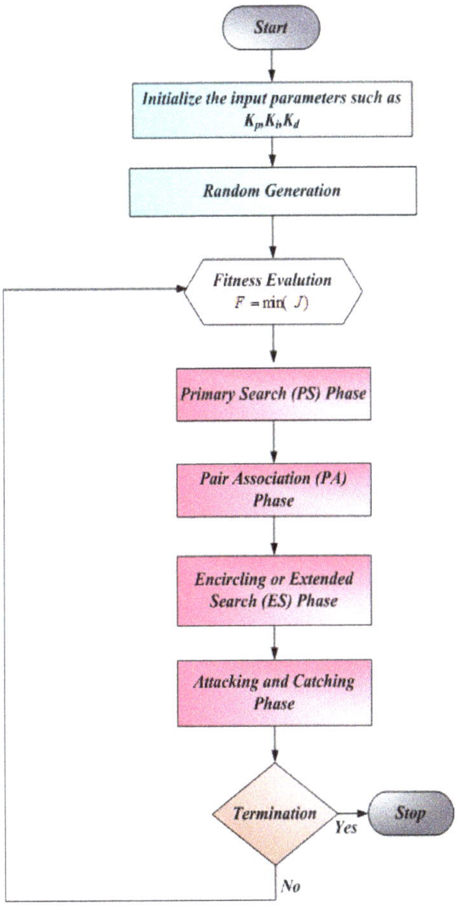

Figure 4. Flowchart of GMEO.

4.2. Dual-Stream Multi-Dependency Graph Neural Network (DMGNN)

In this section, the prediction using a Dual-Stream Multi-Dependency Graph Neural Network (DMGNN) is discussed [31]. DMGNN enhances optimization by efficiently predicting and adjusting PID parameters, capturing complex dependencies in dynamic systems. It was chosen for its ability to model both global and local patterns through its dual-stream architecture, making it well-suited for optimizing PID parameters in nonlinear and time-sensitive systems like buck-boost converters. This capability ensures faster convergence and significantly improves performance.

$$GCN(X, A) = \delta(\overline{D}^{-\frac{1}{2}}\overline{AD}^{-\frac{1}{2}}XW) \tag{24}$$

where the trainable weight matrix for feature transformation is specified by $\overline{D}_{ij} = \sum_j A_{ij}$ and W.

DMGNN captures the complex dependencies and relationships between the PID parameters and the system's dynamic behavior. Its dual-stream architecture learns both global patterns (long-range dependencies) and local patterns (short-range dependencies) between the system states, improving PID parameter adjustment.

$$AW_s = (sigm(W_1F + b_1) \cdot \tanh(W_2F + b_2))W_3 + b_3 \tag{25}$$

Here, the learnable transformation matrix and bias is specified by W and b, correspondingly, while $sigm$ indicates the $sigmoid$ function. The DMGNN refines the parameters K_p, K_d, and K_i by adjusting them locally, ensuring that the PID controller performs optimally under different load conditions and varying operational environments.

$$MAM = \alpha + (1 - \alpha) \cdot sigm(AW_s) \times sigm(AW_s)^T \tag{26}$$

$$F = softmax(\beta\frac{AFM \times AW_s}{\sqrt{d_k}} + (1 - \beta)AW_s) \times [F_1, F_2] \tag{27}$$

where a pre-defined hyper-parameter and reduction to 0 as the training proceeds is specified by α, β is the pre-defined hyper parameter, the features from the two branches are specified by $F_1 = FUB(F)$ and $F_2 = GAB(F)$, and the concatenation operation is specified by $[,]$. The adjusted PID parameters are used to control the buck-boost converter, ensuring that the system's output voltage is maintained at the desired set point while adapting dynamically to load variations.

$$f_{surv}(y_j, F_j) = \prod_{i=0}^{y_j}(1 - f_{hazard}(i, F_j)) \tag{28}$$

where F and y_j variables related to the system's performance and state, respectively, while f_{hazard} represents the computation process of hazard rates. DMGNN dynamically adjusts the ideal PID parameters K_d, K_p, and K_i based on system conditions by learning both global and local patterns in the system's behavior.

5. Result and Discussion

The proposed method's performance is demonstrated in this section using the findings of the simulation. The GMEO-DMGNN approach for buck-boost converter PID controller optimization was proposed in this study. Using the MATLAB R2022b environment, the proposed method is assessed and contrasted with other current methods. The proposed method's simulation is shown below. All simulations begin with an initial capacitor voltage of 12 V to simulate a realistic steady-state start-up scenario rather than zero-state initialization.

Figure 5 depicts the voltage response over time using the IGJO-Optimized PID controller. The voltage reaches a peak of approximately 13.6 V, overshooting the setpoint by around 1.4 V. It then takes about 1000 ms to settle within a 0.2 V band around the setpoint. The larger overshoot and prolonged oscillations highlight a key drawback, leading to a less stable and slower response. Figure 6 depicts the voltage response over time using the ISCA-Optimized PID controller. The voltage rapidly increases, reaching a peak of approximately 13.2 V within the first second, with an overshoot of 1.2 V. The system then settles in about 500 ms; although the settling time is relatively high, the overshoot of 1.2 V suggests that further optimization could improve system stability and reduce peak voltage. Figure 7 depicts the voltage response over time using the CTA-Optimized PID controller. The initial voltage rise shows a peak of 13.6 V, with a substantial overshoot of 1.6 V above the set point. After the peak, the voltage begins to oscillate but stabilizes within approximately 200 ms. However, the significant overshoot and the time taken to stabilize highlight a drawback of the CTA-Optimized PID system, as it results in slower settling and takes longer to reach steady-state performance.

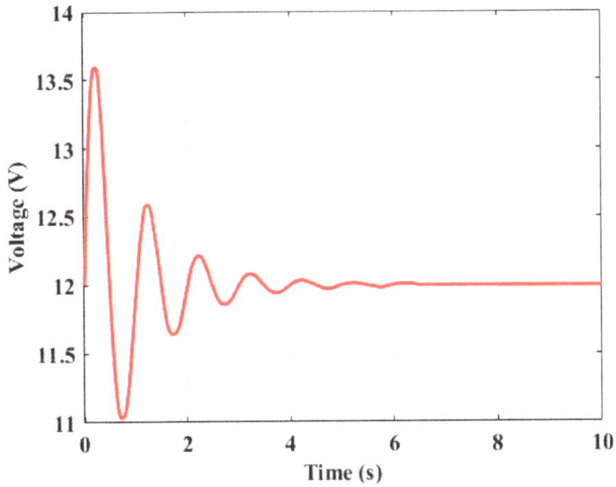

Figure 5. Performance evaluation of the IGJO-Optimized PID controller.

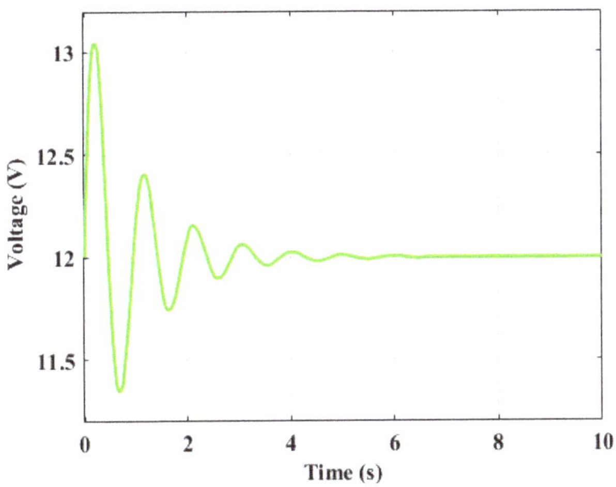

Figure 6. Performance evaluation of the ISCA-Optimized PID controller.

Figure 8 illustrates the voltage response over time using the L-Based Adaptive PID. Initially, the voltage rises quickly, peaking at around 13.3 V, with an overshoot of 1.3 V above the set point. After the peak, the voltage starts to oscillate, but these oscillations dampen over a period of about 150 ms, allowing the system to settle around the set point, with the persistence of oscillations leading to a slower stabilization process that takes several seconds to fully settle.

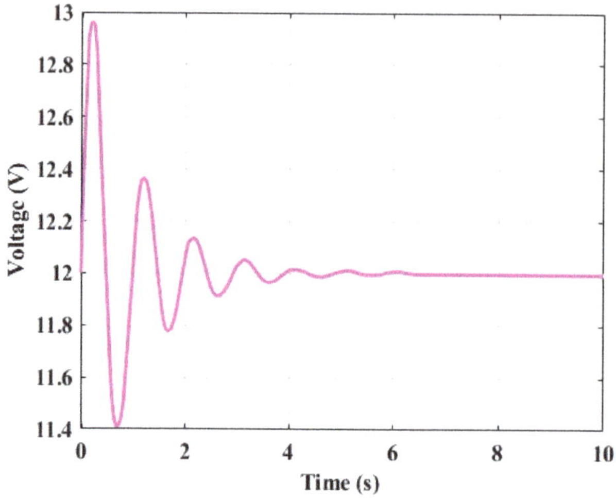

Figure 7. Performance evaluation of CTA-Optimized PID controller.

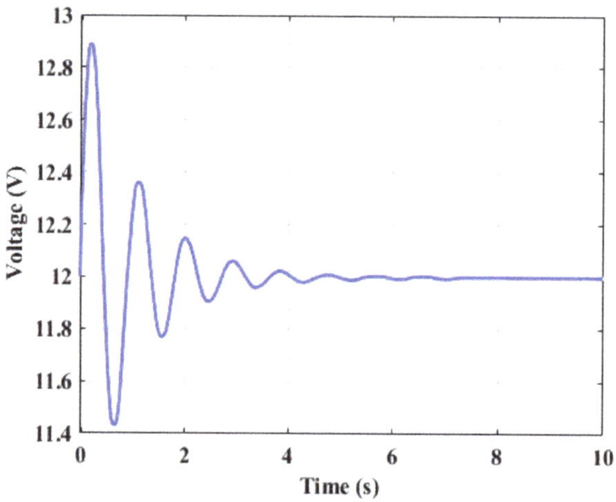

Figure 8. Performance evaluation of L-Based Adaptive PID.

Figure 9 depicts the performance comparison of different control strategies. It compares the performance of various PID control strategies, including IGJO-Optimized PID, ISCA-Optimized PID, CTA-Optimized PID, L-Based Adaptive PID, and the proposed GMEO-DMGNN method. Each strategy shows an initial overshoot followed by oscillations before stabilizing. The IGJO-Optimized PID controller reaches a peak of approximately 13.6 V, overshooting the set point by around 1.4 V, and takes about 1000 ms to settle within a 0.2 V band around the set point. ISCA-Optimized PID reaches a peak of 13.2 V, with a similar overshoot of 1.2 V, and settles in about 500 ms seconds. CTA-Optimized PID shows a peak of 13.6 V, with a substantial overshoot of 1.6 V, and takes approximately 200 ms to stabilize. L-Based Adaptive PID peaks at around 13.3 V with an overshoot of 1.3 V, settling in about 150 ms. The proposed GMEO-DMGNN method shows a peak of 12.9 V, with a more controlled overshoot of 1.1 V, and settles within approximately 100 ms, demonstrating faster convergence and less oscillation. Therefore, the GMEO-DMGNN method outperforms the other techniques, providing a more stable, faster, and more efficient response for controlling buck-boost converters.

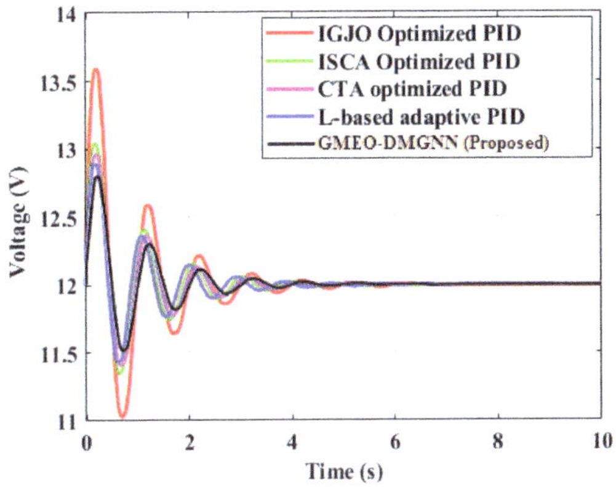

Figure 9. Performance comparison of various control strategies.

Figure 10 depicts the dynamic response characteristics of various control algorithms under different input signals of step and ramp. In subplot (a), under the step input condition, the proposed GMEO-DMGNN method achieves a settling time of 100 ms, an overshoot of only 1.5%, and an extremely low steady-state error of 0.005 V. These values clearly outperform all other compared algorithms. This superior transient response demonstrates GMEO-DMGNN's ability to stabilize the system significantly faster, with minimal overshoot and highly accurate voltage regulation. The improvement is due to the integration of the GMEO optimization algorithm with the DMGNN model, which allows the system to identify optimal PID gains and adaptively refine them based on system behavior. In subplot (b), under the ramp input condition, the GMEO-DMGNN controller maintains a smooth and accurate tracking response with error oscillations confined within ±20, showing stable behavior and precise adaptation to gradual input changes. The smoother slope and absence of sudden fluctuations highlight the method's ability to track dynamic inputs effectively, confirming its superior adaptability over other algorithms.

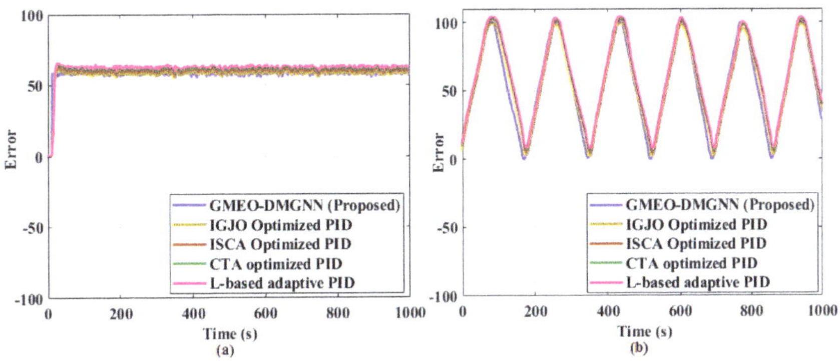

Figure 10. Dynamic response characteristics of various control algorithms under different input signals: (**a**) step input and (**b**) ramp input.

Figure 11 depicts a comparative analysis of the anti-interference performance of various algorithms based on MSE under different iteration counts, indicating convergence behavior and different SNRs. In subplot Figure 11a, the proposed GMEO-DMGNN method demonstrates superior convergence behavior with a significantly lower MSE of 0.020 at 100 iterations compared to IGJO's 0.060, highlighting its faster and more precise optimization capability. This indicates that GMEO-DMGNN consistently outperforms other methods in minimizing error across iterations. In subplot Figure 11b, the proposed GMEO-DMGNN method demonstrates superior anti-interference performance under varying noise conditions. Even at a low SNR of 0 dB, it maintains a low MSE of 0.050. This indicates its strong noise resilience. Compared to algorithms like ISCA, which records a high MSE of 0.185 at 0 dB, GMEO-DMGNN shows significantly better robustness and stability, effectively handling signal disturbances with minimal performance degradation.

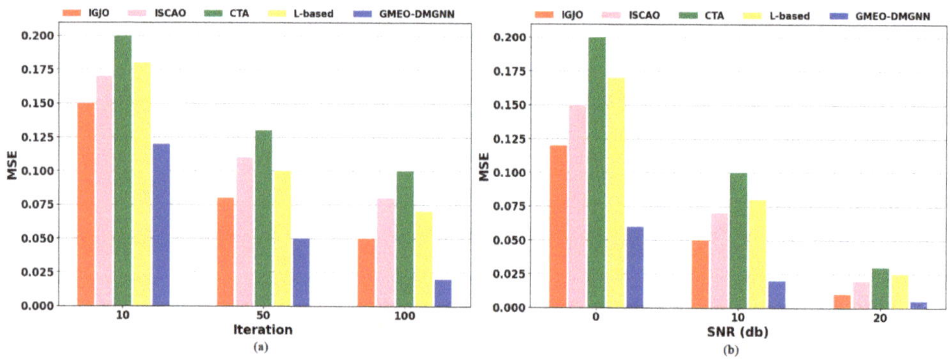

Figure 11. Comparative analysis of the anti-interference performance of various algorithms based on MSE under (**a**) different iteration counts, indicating convergence behavior, and (**b**) different Signal-to-Noise Ratio (SNR).

5.1. Case Studies

Case 1: Trajectory Tracking Analysis

Figure 12 illustrates the trajectory tracking performance of the GMEO-DMGNN under a complex time-varying voltage. Despite rapid changes in the input, the controller accurately tracks the trajectory, with the voltage stabilizing at approximately 12.5 V by 8 s. The peak overshoot reaches close to 13.8 V around 1 s, yet the system quickly compensates and settles without prolonged oscillations. This performance confirms the controller's capability to manage dynamic voltage references with minimal error and fast recovery, thereby validating the effectiveness of the proposed method in complex trajectory tracking scenarios beyond simple step inputs.

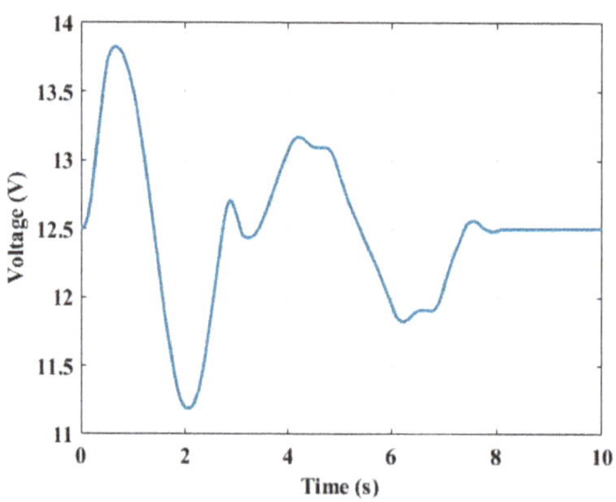

Figure 12. Trajectory tracking performance of the GMEO-DMGNN under a complex time-varying voltage.

Case 2: Disturbance Analysis

Figure 13 depicts the disturbance rejection response when a sudden disturbance is introduced at the converter output. The graph shows the output response of the converter when a disturbance is injected at t = 6 s. Before the disturbance, the system maintains a steady amplitude of approximately 12.5 V. This indicates stable operation under normal conditions. After the disturbance, the GMEO-DMGNN-based PID controller quickly recovers the output with minimal overshoot and fast settling, demonstrating strong disturbance rejection capability.

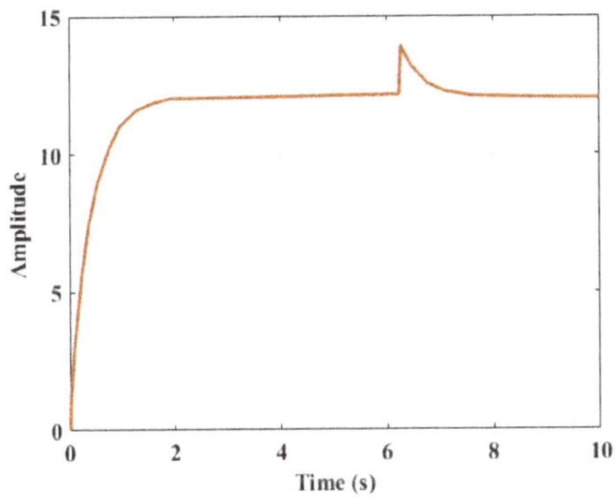

Figure 13. Disturbance rejection response when a sudden disturbance is introduced at the converter output.

Case 3: Uncertainty Analysis

Figure 14 presents the output voltage response of the converter under parameter uncertainty conditions, where some nominal model parameters were deliberately varied. At around t = 2 s and t = 6 s, noticeable disturbances occur due to the introduced parameter changes. The system shows voltage deviations confined within ±0.5 V from the nominal value of 12.5 V when model parameters deviate from their nominal values. Despite the uncertainty, the controller demonstrates a quick recovery, stabilizing the output back to steady state within approximately 1 s after each disturbance. These results confirm the proposed controller's strong robustness and adaptability in the presence of parameter uncertainties, ensuring stable and reliable system performance.

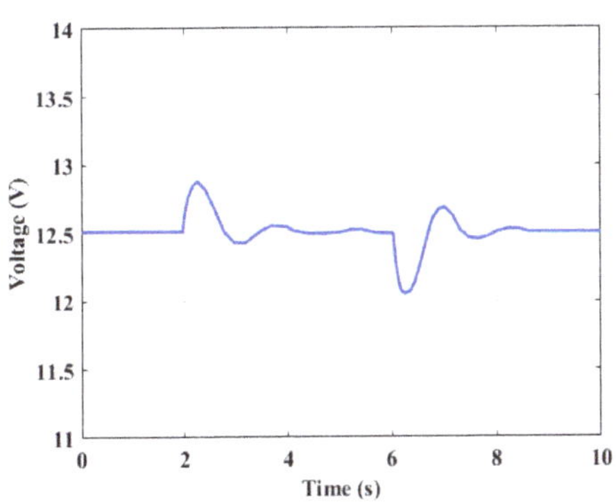

Figure 14. Output voltage response of the proposed controller under parameter uncertainty.

Table 4 compares the performance of various PID optimization methods based on settling time, overshoot, and steady-state error. The IGJO-Optimized PID method has the longest settling time at 1000 ms, with a significant overshoot of 25% and a steady-state error of 0.15 V, indicating slower stabilization and moderate accuracy. With a settling time of 500 ms, a 15% decrease in overshoot, and a lower steady-state error of 0.05 V, the ISCA Optimised PID approach performs better in terms of stability and accuracy. The CTA-Optimized PID method performs even better, with a settling time of 200 ms, a further reduced overshoot of 5%, and a steady-state error of 0.02 V, showing faster convergence and enhanced precision. The L-Based Adaptive PID method offers the fastest settling time at 150 ms, with a minimal overshoot of 3% and an almost negligible steady-state error of 0.01 V, making it highly stable and precise. Finally, the GMEO-DMGNN (Proposed) method outperforms all others with the fastest settling time of 100 ms, the smallest overshoot of 1.5%, and the lowest steady-state error of 0.005 V, providing superior performance in terms of speed, accuracy, and stability. Therefore, the GMEO-DMGNN method stands out as the most efficient and reliable optimization technique for PID controllers.

Table 4. Performance comparison of PID tuning methods.

Methods	Settling Time (ms)	Overshoot (%)	Steady-State Error (V)
IGJO-Optimized PID	1000 ms	25%	0.15 V
ISCA-Optimized PID	500 ms	15%	0.05 V
CTA-Optimized PID	200 ms	5%	0.02 V
L-Based Adaptive PID	150 ms	3%	0.01 V
GMEO-DMGNN (Proposed)	100 ms	1.5%	0.005 V

Table 5 presents a comparison of statistical values between the proposed and existing methods. The proposed method achieves the highest mean value of 4.9876×10^{-7} and the highest standard deviation (SD) of 6.9452×10^{-12}, surpassing the performance of the existing systems. This indicates that the proposed method offers superior variability and greater adaptability in comparison.

Table 5. The statistical performance comparison of proposed with existing techniques.

Methods	Mean	Standard Deviation
IGJO-Optimized PID	4.9876×10^{-7}	6.9452×10^{-12}
ISCA-Optimized PID	3.2344×10^{-7}	5.4322×10^{-12}
CTA-Optimized PID	3.9087×10^{-7}	4.3322×10^{-12}
L-Based Adaptive PID	3.2890×10^{-7}	5.9880×10^{-12}
GMEO-DMGNN (Proposed)	4.0345×10^{-7}	5.4333×10^{-12}

Table 6 compares ITAE and ISE. The proposed system achieves an ITAE of $2.09998998 \times 10^{-13}$ and an ISE of $1.19052756 \times 10^{-6}$. In comparison, existing methods show higher ITAE and ISE values, with the IGJO-Optimized PID achieving an ITAE of $2.39850000 \times 10^{-13}$ and an ISE of $2.73120000 \times 10^{-6}$, the ISCA-Optimized PID yielding an ITAE of $2.36580000 \times 10^{-13}$ and an ISE of $2.62490000 \times 10^{-6}$, the CTA-Optimized PID showing an ITAE of $2.96550000 \times 10^{-13}$ and an ISE of $3.52110000 \times 10^{-6}$, and the L-Based Adaptive PID resulting in an ITAE of 2.9×10^{-13} and an ISE of $2.43120000 \times 10^{-6}$. These results highlight the enhanced accuracy and prediction capability of the GMEO-DMGNN method compared to other systems.

Table 6. Comparison of Time-Weighted Absolute Error (ITAE) and Integral of Squared Error (ISE).

Methods	ITAE	ISE
IGJO-Optimized PID	$2.39850000 \times 10^{-13}$	$2.73120000 \times 10^{-6}$
ISCA-Optimized PID	$2.36580000 \times 10^{-13}$	$2.62490000 \times 10^{-6}$
CTA-Optimized PID	$2.96550000 \times 10^{-13}$	$3.52110000 \times 10^{-6}$
L-Based Adaptive PID	2.9×10^{-13}	$2.43120000 \times 10^{-6}$
GMEO-DMGNN (Proposed)	$2.09998998 \times 10^{-13}$	$1.19052756 \times 10^{-6}$

A comparison of efficiency between the proposed and existing methods is displayed in Table 7. The proposed method achieves the highest efficiency at 98.5%, showcasing its ability to deliver optimal solutions quickly and effectively. The IGJO-Optimized PID method follows with an efficiency of 82%, indicating slightly lower effectiveness. The ISCA-Optimized PID method achieves an efficiency of 89%, while the CTA-Optimized PID method has an efficiency of 89.6%. The L-Based Adaptive PID reaches an efficiency of 91%. Overall, the proposed method stands out for its superior efficiency, making it the most effective solution for optimal performance.

Table 7. Comparison of the efficiency of proposed and existing techniques.

Methods	Efficiency (%)
IGJO-Optimized PID	82%
ISCA-Optimized PID	89%
CTA-Optimized PID	89.6%
L-Based Adaptive PID	91%
GMEO-DMGNN (Proposed)	98.5%

Table 8 compares the optimal PID controller parameters K_p, K_i, K_d for different methods. The proposed GMEO-DMGNN method has the highest K_p at 1.4, K_i at 1001, and a relatively higher K_d at 0.0001, making it more responsive and adaptive compared to existing methods. The way each algorithm tunes these parameters varies based on its internal mechanism. IGJO adjusts gains using Levy flight-based exploration, enabling wider search jumps and resulting in moderately high K_p and K_i but conservative K_d. ISCA uses sinusoidal chaotic sequences for better diversity, which yields slightly lower K_p and K_d. CTA introduces a threshold-aware mutation to refine tuning, leading to balanced gains with a focus on higher K_d damping. The L-based adaptive method modifies gains based on error trends prioritizing K_i for steady-state accuracy. GMEO-DMGNN dynamically learns and adapts PID gains by combining evolutionary search with graph-based learning of system behavior over time, which enables the most effective and balanced tuning among all methods. This makes the GMEO-DMGNN approach more robust and better suited for controlling nonlinear dynamics in the buck-boost converter.

Table 8. Optimal parameters of the PID controller.

Methods	Optimal K_p	Optimal K_i	Optimal K_d
IGJO-Optimized PID	0.89	1766	0.0004
ISCA-Optimized PID	0.98	1733	0.0007
CTA-Optimized PID	0.67	1266	0.0003

Table 8. Cont.

Methods	Optimal K_p	Optimal K_i	Optimal K_d
L-Based Adaptive PID	0.88	1233	0.0009
GMEO-DMGNN (Proposed)	1.4	1001	0.0001

5.2. Discussion

The GMEO-DMGNN approach focuses on optimizing PID controller parameters in buck-boost converters, enhancing their performance under dynamic load conditions. By combining the global optimization capabilities of GMEO with the localized adjustment strengths of DMGNN, the approach effectively fine-tunes PID controller parameters for improved performance in buck-boost converters. The integration of both methods ensures more accurate optimization of PID parameters, enhancing system stability, responsiveness, and overall performance under dynamic conditions. This combined framework offers a more robust and efficient solution for optimizing PID parameters, resulting in improved adaptability, faster convergence, and better control performance in buck-boost converters. The proposed method consistently outperformed existing PID optimization techniques, achieving the fastest settling time of 100 ms, the lowest overshoot of 1.5%, minimal steady-state error of 0.005 V, and the highest efficiency at 98.5%. It also demonstrated superior statistical values with a mean of 4.9876×10^{-7} and a standard deviation of 6.9452×10^{-12}, along with the lowest error integrals, including an ITAE of $2.09998998 \times 10^{-13}$ and an ISE of $1.19052756 \times 10^{-6}$. This result demonstrates that the proposed GMEO-DMGNN system provides superior performance, exhibiting enhanced speed, accuracy, stability, and efficiency compared to existing PID optimization techniques. This finding indicates the GMEO-DMGNN method's comprehensive superiority in PID optimization, achieved through a combination of optimized performance metrics and enhanced hardware utilization, specifically the highest capacitance. Despite its strong performance, the GMEO-DMGNN method may face a few practical limitations. While simulation results are promising, the lack of real-time hardware validation limits immediate deployment insights. The method's reliance on simulated conditions may also result in issues when exposed to unforeseen physical disturbances or sensor noise not captured in the modeling environment. However, these drawbacks can be addressed with further development and real-world testing.

6. Conclusions

The proposed GMEO-DMGNN approach has been introduced to address the limitations of traditional PID controllers in buck-boost converters. In this study, the GMEO-DMGNN combines the global search capabilities of GMEO with the local adjustment strengths of DMGNN to optimize PID controller parameters. Additionally, this method is highly adaptable to dynamic load conditions, maintaining consistent performance across varying operational environments. Unlike conventional methods that may struggle with nonlinearities or varying conditions, the GMEO-DMGNN approach provides enhanced robustness and adaptability. The proposed approach exhibits better performance, achieving a settling time of 100 ms, 1.5% overshoot, and a steady-state error of 0.005 V. This approach outperforms existing methods by offering better adaptability, stability, and efficiency in dynamic and nonlinear environments, making it a robust solution for optimizing PID controllers in buck-boost converters. It also achieves a high efficiency of 98.5% ensuring optimal performance while maintaining system stability under varying conditions. The proposed system reduces the ITAE to $2.09998998 \times 10^{-13}$ and ISE to $1.19052756 \times 10^{-6}$, highlighting the proposed system's superior efficiency in reducing error and improving control performance. The proposed GMEO-DMGNN approach significantly enhances

PID controller performance in buck-boost converters, improving stability, efficiency, and adaptability under dynamic and nonlinear conditions. Future work should prioritize strategies to enhance real-world implementation and reduce computational demands, ensuring that the optimization process remains efficient and practical for practical applications in power electronics.

Author Contributions: Conceptualization, R.B.D. and M.D.; methodology, R.B.D.; software, V.H.K.; validation, R.B.D., M.D. and A.N.; formal analysis, R.B.D.; investigation, R.B.D.; resources, R.B.D.; data curation, R.B.D.; writing—original draft preparation, R.B.D.; writing—review and editing, A.N.; visualization, R.B.D.; supervision, A.N.; project administration, V.H.K.; funding acquisition, M.D. All authors have read and agreed to the published version of the manuscript.

Funding: This research received no external funding.

Data Availability Statement: The original contributions presented in the study are included in the article, further inquiries can be directed to the corresponding author.

Conflicts of Interest: The authors declare no conflict of interest.

References

1. Esfandiari, S.; Davari, M.; Gao, W.; Yang, Y.; Al-Haddad, K. A Novel Converter-Based PV Emulator Control Using Lambert W Method and Fractional-Order Fuzzy Proportional-Integral Controller Trained by Harris Hawks Optimization. *IEEE J. Emerg. Sel. Top. Ind. Electron.* **2024**, *5*, 1493–1507. [CrossRef]
2. Mughees, A.; Mughees, N.; Mughees, A.; Mohsin, S.A.; Ejsmont, K. Enhancing stability and position control of a constrained magnetic levitation system through optimal fractional-order PID controller. *Alex. Eng. J.* **2024**, *107*, 730–746. [CrossRef]
3. Abdolahi, M.; Adabi, J.; Mousavi, S.Y.M. An adaptive extended Kalman filter with passivity-based control for DC-DC converter in DC microgrids supplying constant power loads. *IEEE Trans. Ind. Electron.* **2023**, *71*, 4873–4882. [CrossRef]
4. Kaplan, O.; Bodur, F.; Ozdemir, M.B. Arbitrary Fixed-Time Sliding Mode Control for Buck Converter with Matched and Mismatched Disturbances Based on Fixed-Time Observer. *IEEE Access* **2025**, *13*, 4582–4596. [CrossRef]
5. Gangula, S.D.; Nizami, T.K.; Udumula, R.R.; Chakravarty, A. Self-learning Controller Design for DC–DC Power Converters with Enhanced Dynamic Performance. *J. Control Autom. Electr. Syst.* **2024**, *35*, 532–547. [CrossRef]
6. Kart, S.; Demir, F.; Kocaarslan, İ.; Genc, N. Increasing PEM fuel cell performance via fuzzy-logic controlled cascaded DC-DC boost converter. *Int. J. Hydrogen Energy* **2024**, *54*, 84–95. [CrossRef]
7. Dey, U.; Veerachary, M. Two-Part Controller Design for Switched-Capacitor Based Buck–Boost Converter. *IEEE Trans. Ind. Electron.* 2024; *in press*. [CrossRef]
8. Gaied, H.; Flah, A.; Kraiem, H.; Prokop, L. A comparison between the Quality of Two level and three levels bidirectional Buck-boost converter using the neural network controller. *IEEE Access* **2024**, *12*, 94323–94336. [CrossRef]
9. Lins, A.W.; Krishnakumar, R. Tuning of PID controller for a PV-fed BLDC motor using PSO and TLBO algorithm. *Appl. Nanosci.* **2023**, *13*, 2911–2934. [CrossRef]
10. Liu, X. Design of CCM boost converter utilizing fractional-order PID and Lyapunov-based PID techniques for PF correction. *Electr. Eng.* **2024**, *107*, 3451–3462. [CrossRef]
11. Acosta-Rodríguez, R.A.; Martinez-Sarmiento, F.H.; Múñoz-Hernandez, G.A.; Portilla-Flores, E.A.; Salcedo-Parra, O.J. Validation of Passivity-Based Control and array PID in High-Power Quadratic Buck Converter through Rapid Prototyping. *IEEE Access* **2024**, *12*, 58288–58316. [CrossRef]
12. Sahin, A.K.; Cavdar, B.; Ayas, M.S. An adaptive fractional controller design for automatic voltage regulator system: Sigmoid-based fractional-order PID controller. *Neural Comput. Appl.* **2024**, *36*, 14409–14431. [CrossRef]
13. Kanungo, A.; Kumar, P.; Gupta, V.; Saxena, N.K. A design an optimized fuzzy adaptive proportional-integral-derivative controller for anti-lock braking systems. *Eng. Appl. Artif. Intell.* **2024**, *133*, 108556. [CrossRef]
14. Yakut, Y.B. A new control algorithm for increasing efficiency of PEM fuel cells–Based boost converter using PI controller with PSO method. *Int. J. Hydrogen Energy* **2024**, *75*, 1–11. [CrossRef]
15. Ghamari, S.; Hajihosseini, M.; Habibi, D.; Aziz, A. Design of an Adaptive Robust PI Controller for DC/DC Boost Converter Using Reinforcement-Learning Technique and Snake Optimization Algorithm. *IEEE Access* **2024**, *12*, 141814–141829. [CrossRef]
16. Rajamony, R.; Wang, S.; Ming, W. Modelling and optimal design of a multifunctional single-stage buck-boost differential inverter. *IEEE Open J. Power Electron.* **2024**, *5*, 1328–1339. [CrossRef]
17. Ashok, B.; Michael, P.A. Integration of cascaded controllers for super-lift Luo converter with buck converter in solar photovoltaic and electric vehicle. *Analog Integr. Circuits Signal Process.* **2024**, *118*, 449–466. [CrossRef]

18. Srivastava, A.; Manas, M.; Dubey, R.K. Optimal electrical vehicle-to-grid integration: Three-phase three-level AC/DC converter with model predictive controller based bidirectional power management scheme. *Electr. Eng.* **2024**, 1–23. [CrossRef]
19. Liu, J.; Wei, T.; Chen, N.; Wu, J.; Xiao, P. Fuzzy logic PID controller with both coefficient and error modifications for digitally-controlled DC–DC switching converters. *J. Electr. Eng. Technol.* **2023**, *18*, 2859–2870. [CrossRef]
20. Liu, W.; Liu, Y.; Wang, A. Design of CCM boost converter using fractional-order PID and neural-network techniques for power factor correction. *J. Eng. Appl. Sci.* **2024**, *71*, 227. [CrossRef]
21. Sangeetha, S.; Revathi, B.S.; Balamurugan, K.; Suresh, G. Performance analysis of buck converter with fractional PID controller using hybrid technique. *Robot. Auton. Syst.* **2023**, *169*, 104515. [CrossRef]
22. Warrier, P.; Shah, P.; Sekhar, R. A Comparative performance evaluation of a complex-order PI controller for DC–DC converters. *Results Control Optim.* **2024**, *15*, 100414. [CrossRef]
23. Nanyan, N.F.; Ahmad, M.A.; Hekimoğlu, B. Optimal pid controller for the dc-dc buck converter using the improved sine cosine algorithm. *Results Control Optim.* **2024**, *14*, 100352. [CrossRef]
24. Fong, L.K.; Islam, M.S.; Ahmad, M.A. Optimized PID Controller of DC-DC Buck Converter based on Archimedes Optimization Algorithm. *Int. J. Robot. Control Syst.* **2023**, *3*, 658–672. [CrossRef]
25. Ghamari, S.M.; Khavari, F.; Mollaee, H. Lyapunov-based adaptive PID controller design for buck converter. *Soft Comput.* **2023**, *27*, 5741–5750. [CrossRef]
26. Sharma, P.; Palwalia, D.K.; Sharma, A.K.; Priyadarshi, N.; Padmanaban, S. Coati optimized FOPID controller for non-isolated DC–DC converters in EV charging application. *IET Power Electron.* **2024**, *17*, 2771–2784. [CrossRef]
27. Acevedo, D.M.; Gil-González, W.; Montoya, O.D.; Restrepo, C.; González-Castaño, C. Adaptive Speed Control for a DC Motor Using DC/DC Converters: An Inverse Optimal Control Approach. *IEEE Access* **2024**, *12*, 154503–154513. [CrossRef]
28. Chouay, Y.; Ouassaid, M. An Enhanced Buck-Boost Converter for Photovoltaic Diagnosis application: Accurate MPP Tracker and IV Tracer. *Sci. Afr.* **2025**, *27*, e02561. [CrossRef]
29. Prag, K.; Woolway, M.; Celik, T. Data-driven model predictive control of DC-to-DC buck-boost converter. *IEEE Access* **2021**, *9*, 101902–101915. [CrossRef]
30. Mansour, N.A.; Saraya, M.S.; Saleh, A.I. Groupers and moray eels (GME) optimization: A nature-inspired metaheuristic algorithm for solving complex engineering problems. *Neural Comput. Appl.* **2025**, *37*, 63–90. [CrossRef]
31. Wang, Z.; Ma, J.; Gao, Q.; Bain, C.; Imoto, S.; Liò, P.; Cai, H.; Chen, H.; Song, J. Dual-stream multi-dependency graph neural network enables precise cancer survival analysis. *Med. Image Anal.* **2024**, *97*, 103252. [CrossRef]

energies

MDPI

Article

Stability Analysis and Controller Optimization of MMC in Standalone Mode

Xingyu Liu [1], Shuguang Song [1,*], Wenzhong Ma [1] and Yusheng Wang [2]

1 College of New Energy, China University of Petroleum (East China), Qingdao 266580, China
2 PetroChina Planning & Engineering Institute, Beijing 100080, China
* Correspondence: shuguang.song@upc.edu.cn

Abstract: The modular multilevel converter (MMC) plays an important role in large-scale renewable energy integration and transmission, and it can also operate in standalone mode, powering AC passive loads. This paper focuses on the impact of load variation on the stability of the MMC. First, the impact of load variation on the MMC transfer function is analyzed in detail using the harmonic state-space (HSS) modeling method. Then, by means of the impedance-based stability analysis method, it is found that the MMC tends to become unstable with the increase in inductive loads. If the controller is not well-designed, the system may fail when loads change. Therefore, the worst-case design is used to guarantee the overall system's stability under all load conditions. The impact of traditional proportional resonant (PR) controller parameters on the system's stability is analyzed, revealing that the stability margin and control performance of the controller are limited. Thus, an improved controller structure with an additional series of compensators is proposed. Extensive simulation results in MATLAB/Simulink R2024a verify the analysis of this work and the effectiveness of the proposed controller.

Keywords: controller design; load variation; modular multilevel converter; stability analysis

1. Introduction

Nowadays, high-voltage direct current systems based on modular multilevel converters (MMC-HVDC) have been widely adopted owing to their advantages, such as low switching frequency, high efficiency, and high performance [1–3]. However, stability issues like wide frequency oscillation have been increasingly emerging. A number of incidents have been reported in practical MMC-HVDC projects, e.g., 20~30 Hz sub-synchronous oscillation (SSO) was reported in the Nan'ao three-terminal MMC-HVDC system in China [4] and a 1270 Hz resonance occurred in the Luxi back-to-back MMC-HVDC project [5]. Therefore, the oscillation issues in the MMC-HVDC system have received a lot of attention.

In order to analyze the oscillation phenomenon, many small-signal modeling methods have been developed, such as the dynamic phasor method [6–8], harmonic state-space (HSS) method [2,9,10], and multi-harmonic linearization method [11,12]. Guo et al. developed a comprehensive dynamic phasor-based small-signal model for MMC systems under unbalanced grid conditions, which considered the internal dynamics of the MMC system as well as the more complex controllers, and revealed the reasons for unstable oscillations [6]. Sakinci et al. developed a generalized dynamic phasor model to reduce the order of the conventional dynamic phasor model using the natural frequency separation of the variations, where the impacts of high-order harmonics on the small-signal stability of the MMC were obtained [8]. The HSS modeling approach was first introduced to characterize the multi-harmonic coupling behavior of the MMC in [2], where the accurate small-signal impedance models were developed. Wu et al. established an HSS-based MMC small-signal impedance model with the frequency-coupling dynamics considered, which was used for analyzing the SSO phenomenon [9]. Xu et al. proposed a complete HSS-based MMC small-signal impedance model considering the impedance of the grid, and the mechanism of

MMC grid impedance coupling was illustrated in detail based on the proposed model [10]. Zhu et al. presented the sequence impedance model of MMC based on the multi-harmonic linearization method and analyzed potential resonance modes, where the impact of the circulating current controller on the MMC impedance and system stability was studied [11]. In ref. [12], a special form of the multi-harmonic linearization method was adopted to reduce the complexity of the sequence impedance modeling of MMC depending on the simpler harmonic spectra.

Moreover, many oscillation suppression solutions have also been proposed to enhance the stability of the MMC, which can be divided into two categories: control parameter optimization and control structure modification. Lyu et al. analyzed the impacts of the proportional gain of the AC voltage controller on the AC-side impedance of MMC, confirming that a larger proportional gain can cause a smaller magnitude of the MMC impedance but hardly affected the resonant frequencies [2]. Huang et al. obtained the optimal controller parameters by traversing all the values within the derived boundaries of the regulator parameters, considering the delay generated by digital control [13]. Wang et al. investigated the effects of different control loop parameters on the hybrid passivity property and proposed a parameter design guideline based on the hybrid passivity criterion [14]. Moreover, many scholars also focus on the controller structure [15–21]. Wang et al. proposed a full power decoupling strategy with a new closed-loop DC-bus current proportional (P) controller for the direct-modulation-based MMC, which improved the stability of the MMC system [15]. Gao et al. presented an improved virtual impedance control method, which was the addition of the virtual resistance-inductance to ameliorate the virtual impedance control that can be equivalent to the parallel connection of resistor and inductor, and it was demonstrated that the proposed method showed better performance compared to the traditional virtual resistance control and the phase lag compensation control method [17]. Guo et al. proposed a supplementary damping controller to mitigate the low-frequency oscillations in the MMC-HVDC system under low inertia conditions by introducing a reactive power compensation signal into the reactive power outer loop control of the MMC [19]. Zhu et al. proposed a damping method with virtual arm impedance by introducing common-mode compensation terms in the insertion index to compensate for the stability defects caused by the circulating current injection methods, reducing the submodule capacitor voltage ripples [21].

Generally, MMC-HVDC systems can operate in grid-connected mode [22,23] and standalone mode [9,16]. In grid-connected mode, the MMC connects directly to a power grid, which can be either stiff or weak. In standalone mode, the MMC can connect to renewable power plants and power other different AC loads. Up to now, the grid-connected mode has been studied extensively. However, the standalone mode has not been examined in detail.

In this operation scenario, the connected loads can be modeled as resistive load R_s and inductive load L_s. Based on existing MMC stability analysis and controller design, the system can achieve stable operation under certain load conditions. However, in practical operation, the loads may vary according to the requirement, which changes the MMC operation points. When the load variation is small, the system can remain stable. However, with the increase in load variation, oscillations may appear, and in severe cases, the MMC system could even fail. How the load variation impacts the MMC internal and external small-signal characteristics still remains unknown. Therefore, it is of great importance to find out the MMC stability evolution trend to assist in designing controller structure and its parameters.

Moreover, the basic proportional resonant (PR) controller is commonly adopted for the MMC in standalone mode to achieve AC voltage control. However, the control degrees of freedom for the PR controller are limited. Given the complex harmonic interactions and small-signal dynamics of the MMC, the traditional PR controller can hardly stabilize the MMC system [9]. Additionally, the transient behavior also shows large voltage overshoot and long settling time, which is unsatisfactory. To improve the stability margin and control performance, the existing controller structure needs to be redesigned with consideration of the MMC internal and external small-signal characteristics.

To address the above issues, in this paper, the time-domain model and frequency-domain model are first established based on the average topology of the MMC [4,10], which can effectively reflect the dynamic information of the sub-synchronous frequency range. It is worth noting that the delay generated by digital control affects the high-frequency range dynamics of the MMC. But this paper focuses on the sub-synchronous frequency range, so the delay effect is not considered. Then, the impact of load variation on the transfer function of internal small-signal perturbation to state variables is analyzed in detail using HSS. It is found that load variation significantly affects the transfer function of internal small-signal perturbation to the AC voltage perturbation response. Furthermore, the AC terminal impedance of the MMC is analyzed. It is shown that AC impedance varies along with the AC load, and the system tends to be unstable with the increase in inductive load. Therefore, the MMC controllers need to be well-designed. Otherwise, the system may fail under certain load conditions. Based on the above analysis, the worst-case design for the MMC controller considering load variation is discussed in this paper, which can ensure overall system stability. In existing literature, the proportional resonant (PR) controller is commonly used in the control loop of the MMC system [2,9,24–26]. However, given that the traditional PR controller has limited control performance and is hard to further improve the stability margin, an improved output voltage controller structure is proposed, which introduces additional series compensators according to the MMC internal small-signal dynamic nature. It is demonstrated that the proposed controller can suppress the resonant peak and decrease the phase differences up to ~110°.

The research highlights of this paper are concluded as follows:

(1) Based on the HSS method, the impact of load variation on the MMC internal (control to state variable transfer function) and external (AC terminal impedance) small-signal characteristics is studied in detail.
(2) The worst-case design for the MMC controller is adopted to ensure the overall system's stability. The control performance of the traditional PR controller is analyzed and proven to be limited.
(3) An improved output voltage controller structure with additional series compensators is proposed according to the actual MMC internal and external small-signal dynamic nature, which improves both the stability margin and control performance.

The remainder of this paper is organized as follows: Section 2 reviews the time-domain mathematical model and HSS-based small-signal model of the MMC. The impact of load variation on the MMC transfer function is examined in Section 3. Then, MMC stability is analyzed in detail in Section 4. Section 5 discusses the limitation of the PR controller and proposes the improved controller structure. Section 6 concludes this paper.

2. Modeling of Modular Multilevel Converter

In this section, the topology and mathematical model of the MMC are reviewed.

2.1. Circuit Configuration

The topology of a three-phase MMC operating in standalone mode is depicted in Figure 1. In each phase of MMC, there are two arms, namely the upper arm and the lower arm. Each arm consists of N submodules (SMs) connected in series. and are the arm inductance and resistance, respectively. The half-bridge SM structure is shown on the right-hand side, which contains a floating capacitor and two switching devices. The induction motor, which is a common industrial load, can be expressed as an inductor in series with a resistor [27]. Thus, the AC load in Figure 1 is represented by an inductor L_s connecting with the resistor R_s.

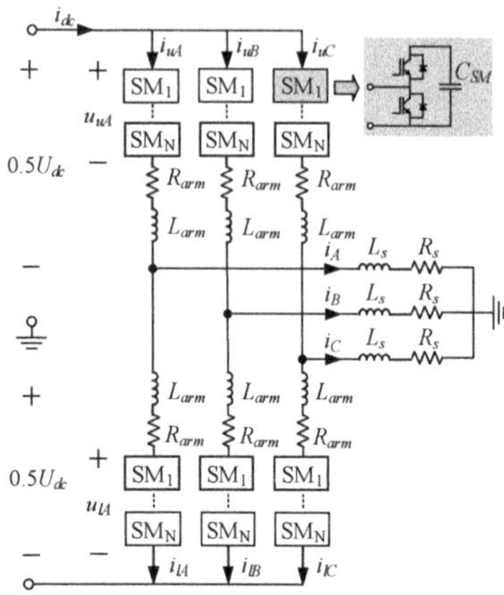

Figure 1. The topology of a three-phase MMC.

For simplification, Figure 2 gives the average model of one phase leg of the MMC, where $v_{cu\Sigma}$ and $v_{cl\Sigma}$ are the sum capacitor voltages for the upper and lower arms, respectively [28]. C_{eq} represents the arm equivalent capacitance, where $C_{eq} = C_{SM}/N$. $m_u(t)$ and $m_l(t)$ are the control signals for upper and lower arms, which can be expressed as [2]

$$\begin{cases} m_u(t) = \frac{1}{2}[1 - m(t) - m_2(t) + \sum_{n=3}^{\infty} m_n(t)] \\ m_l(t) = \frac{1}{2}[1 + m(t) - m_2(t) + \sum_{n=3}^{\infty} m_n(t)] \end{cases} \tag{1}$$

where $m(t)$ and $m_2(t)$ denote the fundamental frequency component for output control and second-order harmonic component for circulating current control, respectively. Moreover, for $n \geq 3$, $m_n(t)$ in (1) is normally equal to zero [2].

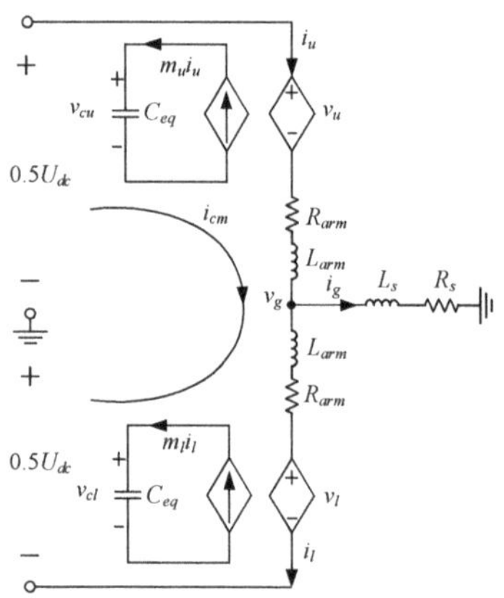

Figure 2. The average model of one phase leg of the MMC.

The circulating current $i_{cm}(t)$ and the AC output current $i_g(t)$ are defined as

$$i_{cm}(t) = \frac{i_u(t) + i_l(t)}{2} \tag{2}$$

$$i_g(t) = i_u(t) - i_l(t) \tag{3}$$

The equivalent arm output voltages can be expressed as

$$v_u(t) = m_u(t)v_{cu\Sigma}(t) \tag{4}$$

$$v_l(t) = m_l(t)v_{cl\Sigma}(t) \tag{5}$$

The capacitor charging and discharging dynamics can be expressed as [9]

$$C_{\text{eq}}\frac{dv_{cu\Sigma}}{dt} = m_u i_u \tag{6}$$

$$C_{eq}\frac{dv_{cl\Sigma}}{dt} = m_l i_l \tag{7}$$

Applying Kirchhoff's law to the upper arm and the lower arm, two circuit equations can be obtained as

$$v_u(t) + R_{arm}i_u(t) + L_{arm}\frac{di_u(t)}{dt} + v_g(t) = \frac{U_{dc}}{2} \tag{8}$$

$$v_l(t) + R_{arm}i_l(t) + L_{arm}\frac{di_l(t)}{dt} - v_g(t) = \frac{U_{dc}}{2} \tag{9}$$

where $v_g(t) = R_s i_g(t) + L_s[di_g(t)/dt]$.

Applying the differential-mode (DM) and common-mode (CM) definitions [29], the sum and difference in the control signals and equivalent SM capacitor voltages for the lower and upper arms in Figure 2 can be defined as

$$m_l(t) + m_u(t) = 2m_{cm}(t) \tag{10}$$

$$m_l(t) - m_u(t) = 2m_{dm}(t) \tag{11}$$

$$v_{cl\Sigma}(t) + v_{cu\Sigma}(t) = 2u_{Ccm}(t) \tag{12}$$

$$v_{cl\Sigma}(t) - v_{cu\Sigma}(t) = 2u_{Cdm}(t) \tag{13}$$

Substituting (1) into (10) and (11), CM and DM control signals can be written as

$$m_{cm}(t) = \frac{1}{2}[1 - m_2\cos(2\omega_1 t + \theta_{m2})] \tag{14}$$

$$m_{dm}(t) = \frac{1}{2}m\cos(\omega_1 t + \theta_{m1}) \tag{15}$$

where ω_1 represents the fundamental angular frequency.

Substituting (2)–(5) and (10)–(13) into (6)–(9), MMC dynamics in DM and CM representation are given as follows:

$$\begin{cases} \frac{di_{cm}(t)}{dt} = -\frac{R_{arm}}{L_{arm}}i_{cm}(t) - \frac{m_{cm}(t)}{L_{arm}}u_{Ccm}(t) - \frac{m_{dm}(t)}{L_{arm}}u_{Cdm}(t) + \frac{1}{2L_{arm}}U_{dc} \\ \frac{di_g(t)}{dt} = -\frac{R_{arm}+2R_s}{L_{arm}+2L_s}i_g(t) + \frac{2m_{dm}(t)}{L_{arm}+2L_s}u_{Ccm}(t) + \frac{2m_{cm}(t)}{L_{arm}+2L_s}u_{Cdm}(t) \\ \frac{du_{Ccm}(t)}{dt} = \frac{m_{cm}(t)}{C_{\text{eq}}}i_{cm}(t) - \frac{m_{dm}(t)}{2C_{\text{eq}}}i_g(t) \\ \frac{du_{Cdm}(t)}{dt} = \frac{m_{dm}(t)}{C_{\text{eq}}}i_{cm}(t) - \frac{m_{cm}(t)}{2C_{\text{eq}}}i_g(t) \end{cases} \tag{16}$$

2.2. HSS-Based MMC Steady-State Model

Based on (16), choosing the circulating current $i_{cm}(t)$, output current $i_g(t)$, CM arm capacitor voltage $u_{Ccm}(t)$, and DM arm capacitor voltage $u_{Cdm}(t)$ as state variables, the time-domain state-space model of the MMC can be expressed as

$$\dot{x}(t) = A(t)x(t) + B(t)u(t) \tag{17}$$

where

$$\dot{x}(t) = [i_{cm}(t), i_g(t), u_{Ccm}(t), u_{Cdm}(t)]^T \tag{18}$$

$$A(t) = \begin{bmatrix} -\frac{R_{arm}}{L_{arm}} & 0 & -\frac{m_{cm}(t)}{L_{arm}} & -\frac{m_{dm}(t)}{L_{arm}} \\ 0 & -\frac{R_{arm}+2R_s}{L_{arm}+2L_s} & \frac{2m_{dm}(t)}{L_{arm}+2L_s} & \frac{2m_{cm}(t)}{L_{arm}+2L_s} \\ \frac{m_{cm}(t)}{C_{eq}} & -\frac{m_{dm}(t)}{2C_{eq}} & 0 & 0 \\ \frac{m_{dm}(t)}{C_{eq}} & -\frac{m_{cm}(t)}{2C_{eq}} & 0 & 0 \end{bmatrix} \tag{19}$$

$$B(t) = diag(\frac{1}{2L_{arm}}, 0, 0, 0) \tag{20}$$

$$u(t) = [U_{dc}, 0, 0, 0]^T \tag{21}$$

The detailed HSS modeling procedure can be found in [2,9]. Thus, it is not repeated in this paper. Based on the Fourier series and harmonic balance theory, (17)–(21) can be transformed into the frequency-domain HSS model, which is expressed as

$$s\mathbf{X} = (\mathbf{A} - \mathbf{Q})\mathbf{X} + \mathbf{BU} \tag{22}$$

where

$$\mathbf{X} = [X_{-h}, \ldots, X_{-1}, X_0, X_1, \ldots, X_h]^T \tag{23}$$

$$\mathbf{A} = \begin{bmatrix} A_0 & A_{-1} & \cdots & A_{-h} & & & \\ A_1 & \ddots & \ddots & \ddots & \ddots & & \\ \ddots & \ddots & A_0 & A_{-1} & \ddots & \ddots & \\ A_h & \ddots & \ddots & A_0 & A_{-1} & \ddots & A_{-h} \\ & \ddots & \ddots & A_1 & A_0 & \ddots & \vdots \\ & & \ddots & \ddots & \ddots & \ddots & A_{-1} \\ & & & A_h & \ddots & A_1 & A_0 \end{bmatrix} \tag{24}$$

$$\mathbf{B} = \begin{bmatrix} B_0 & B_{-1} & \cdots & B_{-h} & & & \\ B_1 & \ddots & \ddots & \ddots & \ddots & & \\ \ddots & \ddots & B_0 & B_{-1} & \ddots & \ddots & \\ B_h & \ddots & \ddots & B_0 & B_{-1} & \ddots & B_{-h} \\ & \ddots & \ddots & B_1 & B_0 & \ddots & \vdots \\ & & \ddots & \ddots & \ddots & \ddots & B_{-1} \\ & & & B_h & \ddots & B_1 & B_0 \end{bmatrix} \tag{25}$$

$$\mathbf{U} = [U_{-h}, \ldots, U_{-1}, U_0, U_1, \ldots, U_h]^T \tag{26}$$

$$\mathbf{Q} = diag(-jh\omega_1\mathbf{I}, \ldots, -j\omega_1\mathbf{I}, \mathbf{0}, j\omega_1\mathbf{I}, \ldots, jh\omega_1\mathbf{I}) \tag{27}$$

where **A** and **B** are called Toeplitz matrix. The subscripts h of the matrix elements represent the harmonic order. For instance, A_h, B_h are the corresponding h th Fourier coefficients of $A(t)$ and $B(t)$, respectively.

Letting the left-hand side of (22) be zero, the steady-state operation point of MMC can be calculated by

$$\mathbf{X}_{ss} = -(\mathbf{A} - \mathbf{Q})^{-1}(\mathbf{BU}) \tag{28}$$

2.3. HSS-Based MMC Small-Signal Model

To derive the small-signal model of the MMC, the harmonic linearization principle needs to be adopted. Applying the small perturbation to (17), the MMC small-signal state-space equation can be obtained as

$$x_p(t) = A(t)x_p(t) + A_p(t)x(t) + B(t)u_p(t) \tag{29}$$

where subscript p denotes the small-signal perturbation components.

In steady state, MMC contains $\pm h$ th ($h \in Z$) steady-state harmonics. According to the operation principle of the MMC, the injected small-signal perturbation with angular frequency ω_p can be coupled with these steady-state harmonics, which generates a series of harmonics with angular frequency $\omega_p \pm h\omega_1$ ($h \in Z$).

Applying the Fourier series and harmonic balance theory, (29) can be transformed into the frequency-domain model as follows:

$$s\mathbf{X}_p = (\mathbf{A} - \mathbf{Q}_p)\mathbf{X}_p + \mathbf{A}_p\mathbf{X} + \mathbf{BU}_p \tag{30}$$

$$\mathbf{X}_p = [X_{p-h}, \cdots, X_{p-1}, X_p, X_{p+1}, \cdots, X_{p+h}]^T \tag{31}$$

$$\mathbf{U}_p = [U_{p-h}, \cdots, U_{p-1}, U_p, U_{p+1}, \cdots, U_{p+h}]^T \tag{32}$$

$$\mathbf{A}_p = \begin{bmatrix} A_p & A_{p-1} & \cdots & A_{p-h} & & & \\ A_{p+1} & \ddots & \ddots & \ddots & \ddots & & \\ \ddots & \ddots & A_p & A_{p-1} & \ddots & \ddots & \\ A_{p+h} & \ddots & \ddots & A_p & A_{p-1} & \ddots & A_{p-h} \\ & \ddots & \ddots & A_{p+1} & A_p & \ddots & \vdots \\ & & \ddots & \ddots & \ddots & \ddots & A_{p-1} \\ & & & A_{p+h} & \ddots & A_{p+1} & A_p \end{bmatrix} \tag{33}$$

$$\mathbf{Q}_p = diag[j(\omega_p - h\omega_1)\mathbf{I}, \cdots, j(\omega_p - \omega_1)\mathbf{I}, j\omega_p\mathbf{I}, j(\omega_p + \omega_1)\mathbf{I}, \cdots, j(\omega_p + h\omega_1)\mathbf{I}] \tag{34}$$

However, in closed-loop control, the small-signal coefficient matrix $\mathbf{A}_p$ is related to the small-signal state variables $\mathbf{X}_p$. Therefore, (30) cannot be calculated directly, and a variable separation is needed [30], where (30) can be further reorganized as

$$s\mathbf{X}_p = (\mathbf{A} - \mathbf{Q}_p - \mathbf{G})\mathbf{X}_p + \mathbf{GR_p} + \mathbf{BU}_p \tag{35}$$

where **G** is the control matrix, which is composed of the product of the controller transfer function and steady-state state variables, and $\mathbf{R_p}$ represents the small-signal perturbation of the control reference.

Letting the left-hand side of (35) be zero, the small-signal perturbation of the state variables $\mathbf{X}_p$ can be calculated by

$$\mathbf{X}_p = -(\mathbf{A} - \mathbf{Q}_p - \mathbf{G})^{-1}(\mathbf{G}\mathbf{R}_\mathbf{p} + \mathbf{B}\mathbf{U}_p) \tag{36}$$

Based on the MMC small-signal model, many scholars established the MMC terminal impedance and transfer function model to investigate the system stability and controller design issues.

3. Impact of Load Variation

In this section, the traditional controller structure is first introduced. Then, the transfer functions of internal perturbation to AC voltage and circulating current are analyzed, respectively. To support the analysis, an MMC simulation system is built in MATLAB/Simulink R2024a. The main parameters of the MMC are given in Table 1. Table 2 shows the different load conditions studied in this paper, where $|Z_s|$ remains the same but R_s/X_s varies.

Table 1. Parameters of the MMC model.

Parameters	Symbols	Values
Rated output power	P_o	100 MVA
Grid frequency	f_g	50 Hz
DC bus voltage	U_{dc}	±160 kV
AC L-L voltage (rms)	V_{grms}	166 kV
Number of the SM per arm	N	200
SM capacitance	C_{SM}	2.5 mF
Equivalent SM capacitance	C_{eq}	12.5 μF
Arm inductance	L_{arm}	0.18 mH
Arm resistance	R_{arm}	4.06 Ω

Table 2. Parameters of the AC load.

Load Conditions	R_s	L_s
Case I	275.56 Ω	0
Case II	194.85 Ω	0.6202 H
Case III	123.23 Ω	0.7845 H
Case IV	87.14 Ω	0.8321 H
Case V	54.04 Ω	0.8601 H
Case VI	45.30 Ω	0.8652 H
Case VII	34.18 Ω	0.8704 H
Case VIII	0	0.8771 H

3.1. Traditional Controller Structure

The MMC always operates in closed-loop mode rather than open-loop mode. Therefore, the traditional controller structure is introduced. In standalone mode, MMC commonly uses the voltage-controlled closed-loop form, where the AC voltage controller structure traditionally adopts the basic proportional resonant (PR) controller [2,9]. Additionally, the circulating current controller is also added in the closed-loop configuration, where the PR controller is used to suppress the second-order circulating current.

Figure 3 shows the block diagrams of the PR controllers for AC voltage and circulating current control. The delay generated by digital control affects the high-frequency range dynamics of MMC [13,31]. However, as this paper focuses on the sub-synchronous frequency range, the delay effect is not considered. The corresponding transfer functions are expressed as

$$G_{vg}(s) = K_{Pvg} + \frac{2K_{Rvg}\omega_c s}{s^2 + 2\omega_c s + \omega_1^2} \tag{37}$$

$$G_{icm}(s) = K_{Picm} + \frac{2K_{Ricm}\omega_c s}{s^2 + 2\omega_c s + (2\omega_1)^2} \tag{38}$$

where K_{Pvg}, K_{Picm}, K_{Rvg}, and K_{Ricm} are the proportional and resonant gain for the AC voltage and circulating current controller, respectively. Moreover, the resonant angular frequencies of the AC voltage and circulating current controller are tuned at ω_1 and $2\omega_1$, respectively.

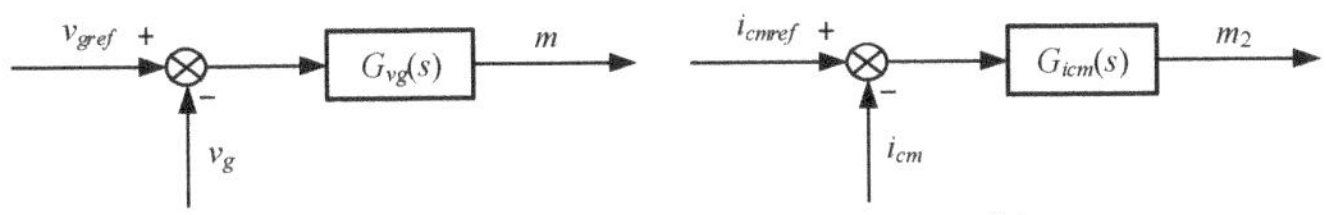

(a) (b)

Figure 3. Block diagrams of the MMC controllers (**a**) AC voltage control; (**b**) circulating current control.

Based on the control theory, open-loop gain is crucial for analyzing the system stability, which is composed of controller and control to state variable transfer function. Therefore, in the following parts, the impact of AC load variation on the control to AC voltage and circulating current transfer functions is studied in detail.

3.2. Transfer Function of Internal Perturbation to AC Voltage

The AC voltage is controlled by the DM control signal. Thus, the internal DM small-signal perturbation $m_{dm,p}(t)$ at angular frequency ω_p with amplitude $M_{dm,p}$ in the MMC control signal should be added, which can be expressed as

$$m_{dm,p}(t) = M_{dm,p}\cos(\omega_p t) \tag{39}$$

The corresponding perturbation matrix consisting of internal DM small-signal perturbation can be expressed as $A_{dm,p}(t)$. Then, the state variable small-signal perturbation response can be expressed by

$$\mathbf{X}_p = -(\mathbf{A} - \mathbf{Q}_p)^{-1}(\mathbf{A}_{dm,p}\mathbf{X}) \tag{40}$$

By solving (40), the corresponding AC current small-signal perturbation response $i_{g,p}(t)$ can be obtained. The value of the injected internal DM small-signal perturbation to AC current transfer function at angular frequency ω_p can be calculated by

$$T_{MMC}(j\omega_p) = \frac{\mathbf{I}_{g,p}}{\mathbf{M}_{dm,p}} \tag{41}$$

where $\mathbf{I}_{g,p}$ and $\mathbf{M}_{dm,p}$ represent the phase of the AC current small-signal perturbation and the injected internal DM small-signal perturbation at angular frequency ω_p, respectively.

The value of AC voltage small-signal perturbation at angular frequency ω_p can be obtained by

$$\mathbf{U}_{g,p} = (R_s + j\omega_p L_s)\mathbf{I}_{g,p} \tag{42}$$

As a result, the value of the injected internal DM small-signal perturbation to the AC voltage transfer function at angular frequency ω_p can be calculated by

$$T_{MMC1}(j\omega_p) = \frac{(R_s + j\omega_p L_s)\mathbf{I}_{g,p}}{\mathbf{M}_{dm,p}} \tag{43}$$

And the transfer function measurement process in simulation can be described as follows: First, a DM small-signal perturbation $m_{dm,p}(t) = M_{dm,p}\cos(\omega_p t)$ is injected at angular frequency ω_p into the control signal. Then, the fast Fourier transform (FFT) module is used to obtain the magnitude and phase of the corresponding AC voltage small-signal response at angular frequency ω_p. Change the perturbation frequency, and repeat the whole process until the measurement needs are met.

Figure 4 shows the bode diagram of the injected internal DM small-signal perturbation to AC voltage transfer functions under different AC load conditions. The solid lines and marks represent the analytical and simulated results of $T_{MMC1}(j\omega_p)$ in (43), respectively. The analytical results clearly agree well with the simulation results. It can be observed that the magnitude response is generally flat at lower frequencies. Then, a resonant peak occurs at ~28 Hz, which is augmented greatly with the increase in inductive load. The magnitude difference can reach up to 19 dB. Correspondingly, the phase response shows a significant shift around the frequency where the magnitude peaks.

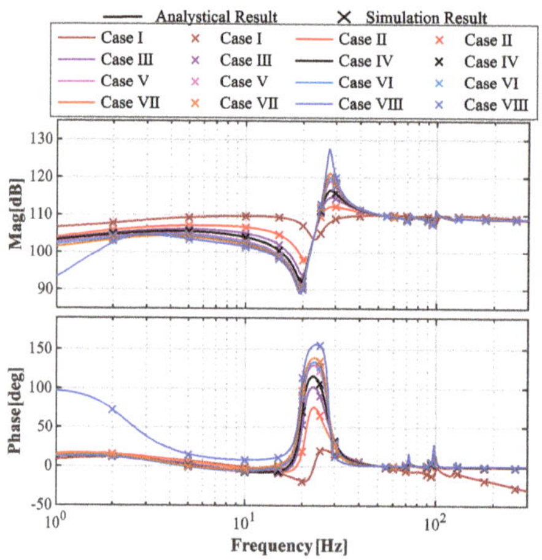

Figure 4. Model verification of internal DM perturbation to AC voltage transfer functions under different load conditions.

3.3. Transfer Function of Internal Perturbation to Circulating Current

The circulating current is controlled by the CM control signal. Therefore, a predefined internal CM small-signal perturbation $m_{cm,p}(t)$ can be injected into the control function, whose expression can be written as

$$m_{cm,p}(t) = M_{cm,p}\cos(\omega_p t) \tag{44}$$

where $M_{cm,p}$ and ω_p are the amplitude and angular frequency of internal CM small-signal perturbation, respectively. Similarly, the corresponding state variables small-signal perturbation can be calculated by

$$\mathbf{X}_p = -(\mathbf{A} - \mathbf{Q}_p)^{-1}(\mathbf{A}_{cm,p}\mathbf{X}) \tag{45}$$

Therefore, the transfer function of internal CM small-signal perturbation to circulating current at angular frequency ω_p can be calculated by

$$T_{MMC2}(j\omega_p) = \frac{\mathbf{I}_{cm,p}}{\mathbf{M}_{cm,p}} \tag{46}$$

The measurement process is similar except that an internal CM small-signal perturbation is injected in simulation. Figure 5 shows the bode diagram of the injected internal CM small-signal perturbation to circulating current transfer functions under different load conditions. The solid lines and marks are the analytical and simulation results of $T_{MMC2}(j\omega_p)$ in (46), respectively. The analytical results agree well with the simulation results as well. Compared with the AC voltage transfer function, both magnitude and phase responses show weak correlation with different load cases. Phase differences can only be observed below 10 Hz, while there is no obvious magnitude difference, indicating that the MMC internal circulating current small-signal dynamics are hardly affected.

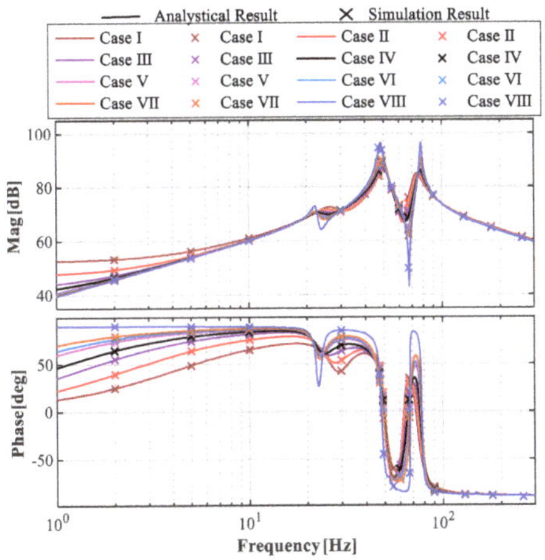

Figure 5. Model verification of internal CM perturbation to circulating current transfer functions under different load conditions.

Based on the above analysis, the load variation has a larger impact on the transfer function of internal DM small-signal perturbation to the AC voltage perturbation response. In other words, it greatly affects the stability of AC voltage control. Therefore, in the next section, a detailed discussion of the MMC stability under different load conditions is carried out.

4. MMC Stability Analysis Considering Load Variation

The analysis in the previous section verifies that load variation has a great impact on MMC small-signal dynamics. In this section, MMC stability is discussed in detail based on the impedance-based stability analysis method.

4.1. Impedance-Based Stability Analysis Method for the MMC System

To analyze the stability of closed-loop systems, the impedance-based stability analysis method is adopted. Figure 6 shows the Thevenin equivalent circuit of the MMC operating in standalone mode, where the current flowing from the MMC system to the load can be written as [32,33]

$$I(s) = \frac{v_{MMC}(s)}{Z_{Load}(s) + Z_{MMC}(s)} \tag{47}$$

which can be reorganized as

$$I(s) = \frac{v_{MMC}(s)}{Z_{Load}(s)} \cdot \frac{1}{1 + Z_{MMC}(s)/Z_{Load}(s)} \tag{48}$$

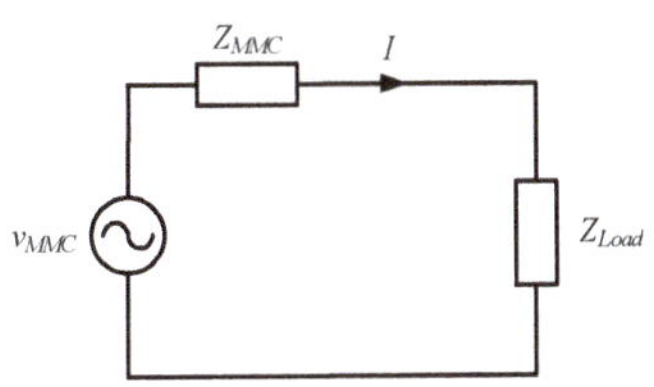

Figure 6. Thevenin equivalent circuit of MMC operating in standalone mode.

It can be assumed that the MMC voltage is stable when unloaded and the load current is stable when powered from an ideal source. Therefore, both $v_{MMC}(s)$ and $1/Z_{Load}(s)$ are stable. Then, current stability depends on (49)

$$H(s) = \frac{1}{1 + Z_{MMC}(s)/Z_{Load}(s)} \tag{49}$$

By linear control theory, $H(s)$ is stable if and only if $Z_{MMC}(s)/Z_{Load}(s)$ satisfies the Nyquist stability criterion, where the details are as follows [34]:

"If and only if the total number of counterclockwise encirclements of $(-1 + j0)$ by the eigen-loci of $Z_{MMC}(s)/Z_{Load}(s)$ is equal to the total number of right-half-plane (RHP) poles of $Z_{MMC}(s)$ and $1/Z_{Load}(s)$, the closed-loop MMC system shown in Figure 6 is stable".

Considering that $Z_{MMC}(s)$ and $1/Z_{Load}(s)$ have no RHP poles, the above criterion can be reorganized as follows:

"If and only if the eigen-loci of $Z_{MMC}(s)/Z_{Load}(s)$ does not encircle $(-1 + j0)$ counterclockwise, the closed-loop MMC system shown in Figure 6 is stable".

4.2. Stability Analysis of the Closed-Loop MMC System Under Different Load Conditions

Figure 7a shows the bode diagram of $Z_{close0}(s)$ and $Z_{Load}(s)$ of load case II with controller parameters in Table 3. The analytical results agree well with the simulation results. It is observed that the phase difference at the intersection point of $Z_{close0}(s)$ and $Z_{Load}(s)$ is much lower than 180°, which satisfies the Nyquist stability criterion. Figure 7b shows the time-domain waveforms of output voltage and current, indicating that the system works stably. Figure 7c shows the time-domain waveforms of circulating current in phase A, which fluctuates around 75 A, indicating stable system operation. And the sum capacitor voltages for the upper and lower arms remain stable, as shown in Figure 7d. It is noted that AC voltage controller parameters in Table 3 are designed based on load case II.

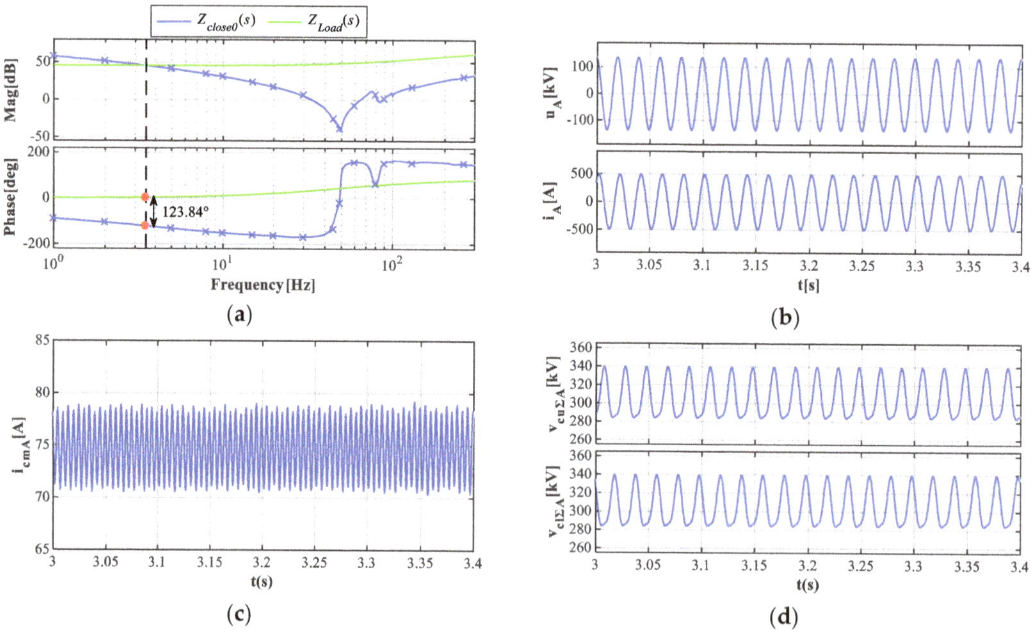

Figure 7. Bode diagrams and time-domain waveforms of load case II. (**a**) $Z_{close0}(s)$ and $Z_{Load}(s)$ (solid lines: analytical results; cross markers: simulation results); (**b**) output voltage and current; (**c**) circulating current; (**d**) sum capacitor voltages for the upper and lower arms.

Table 3. Parameters of the traditional AC voltage PR controller.

Controller Parameters	K_{Pvg}	K_{Rvg}	ω_c
Values	0.8	900	π

Figure 8a shows the bode diagram of $Z_{close0}(s)$ and $Z_{Load}(s)$ of load case VI. It can be seen that the phase difference at the intersection point is higher than 180°, which dissatisfies the Nyquist stability criterion. Figure 8b shows the corresponding time-domain waveforms of output voltage and current, where harmonic components appear. As shown in Figure 8c, the circulating current in phase A exhibits distortion. In Figure 8d, the sum capacitor voltages for the upper and lower arms contain harmonic components.

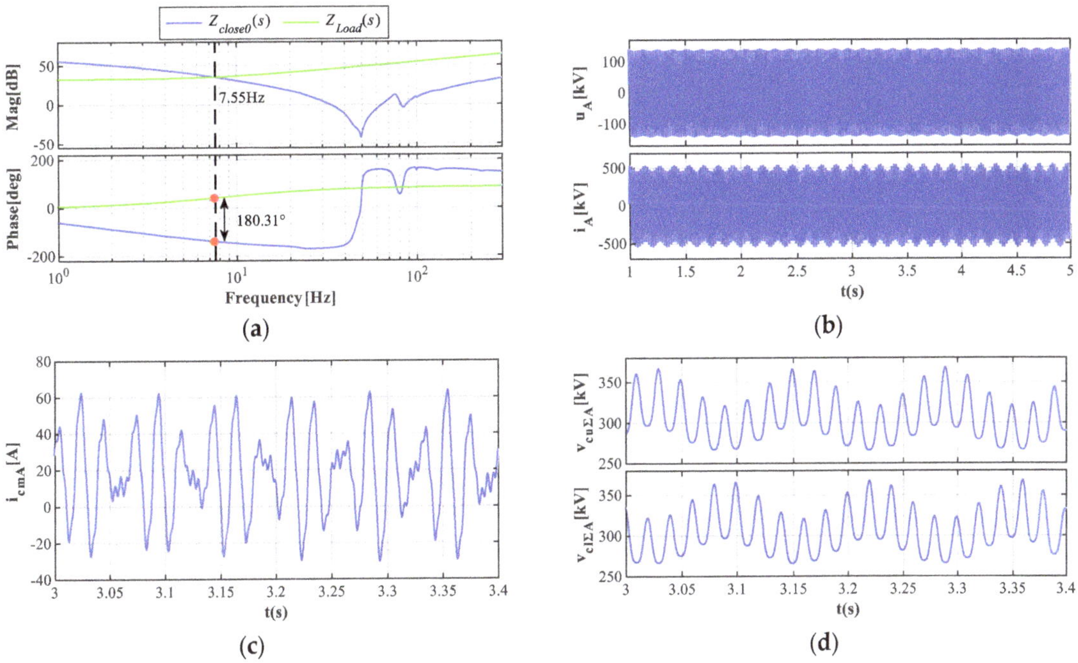

Figure 8. Bode diagrams and time-domain waveforms of load case VI. (**a**) $Z_{close0}(s)$ and $Z_{Load}(s)$; (**b**) output voltage and current; (**c**) circulating current; (**d**) sum capacitor voltages for the upper and lower arms.

Figure 9a shows the bode diagram of $Z_{close0}(s)$ and $Z_{Load}(s)$ of load case VII. It is observed that the phase difference at the intersection point is higher than 180°, which indicates the system dissatisfies the Nyquist stability criterion. Figure 9b shows the corresponding time-domain waveforms of output voltage and current, which contain large harmonic components, confirming the analysis. The corresponding circulating current contains harmonic components, as shown in Figure 9c. Similarly, in Figure 9d, harmonic components also appear in the sum capacitor voltages for the upper and lower arms. The waveforms in Figure 9c,d have already become distorted. Furthermore, time-domain waveforms of load case VIII are not shown because the system is severely unstable.

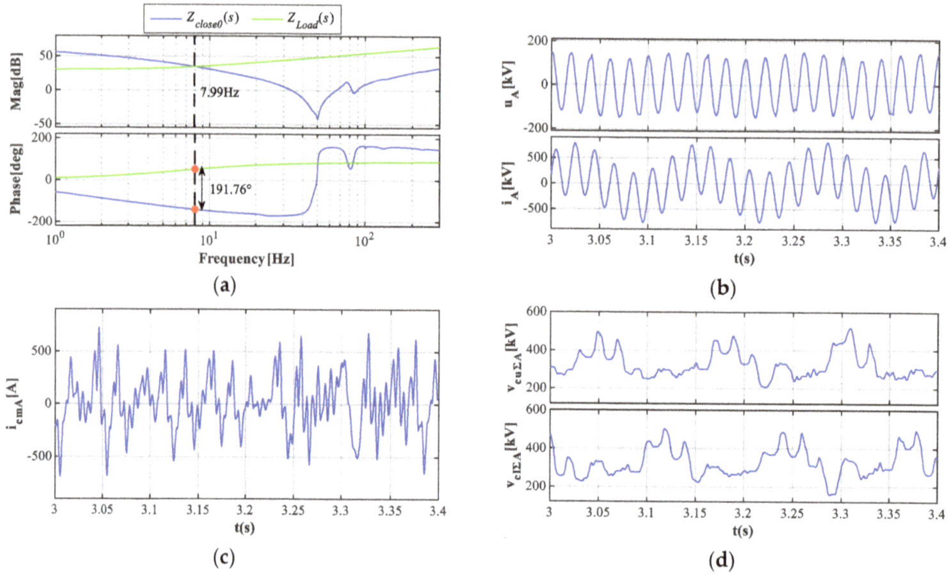

Figure 9. Bode diagrams and time-domain waveforms of load case VII. (**a**) $Z_{close0}(s)$ and $Z_{Load}(s)$; (**b**) output voltage and current; (**c**) circulating current; (**d**) sum capacitor voltages for the upper and lower arms.

Additionally, Figure 10 shows the Fourier analysis results of Figures 8b and 9b. In Figure 10a, it can be seen that SSO appears in the AC output voltage and current, where the resonant frequency is 7.5 Hz, which is close to the analytical resonant frequency of 7.55 Hz obtained in Figure 8a. In Figure 10b, SSO of 7.9 Hz appears in the AC output voltage and current. The resonant frequency of the simulation is close to the analytical resonant frequency of 7.99 Hz obtained in Figure 9a.

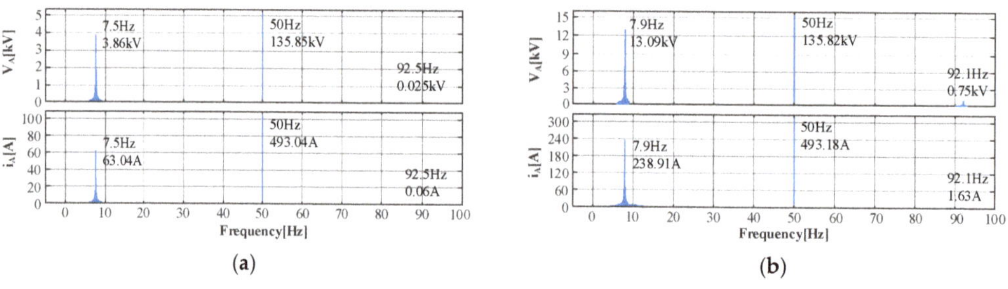

Figure 10. Fourier analysis results of load cases VI and VII. (**a**) Load case VI; (**b**) load case VII.

Based on the transfer functions shown in Figure 4 and the analysis above, the stability of the MMC with the same AC voltage controller under different load conditions is varying. The system tends to be unstable with the increase in inductive load. Thus, to guarantee the overall stability, the worst-case design is adopted for the MMC AC voltage controller design.

4.3. Verification of the Worst-Case Design

The worst load condition is case VIII. Therefore, controller parameters in Table 3 need to be adjusted according to its transfer function feature. The resonant gain K_{Rvg} is changed to 30. Figure 11 shows the bode diagrams of $Z_{close0}(s)$ and $Z_{Load}(s)$ with all load conditions. It can be seen that the phase difference at the intersection point of $Z_{close0}(s)$ and $Z_{Load}(s)$

in the worst case, VIII, is largest but lower than 180°. The phase differences are generally reduced with the decrease in inductive load, and all load conditions satisfy the Nyquist stability criterion, which ensures the overall system stability.

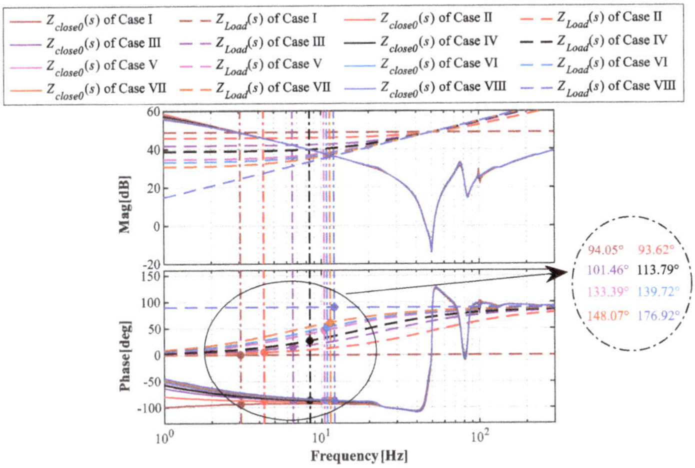

Figure 11. Bode diagrams of $Z_{close0}(s)$ and $Z_{Load}(s)$ under different load conditions.

5. Controller Optimization

5.1. Controller Parameters Optimization

By changing controller parameters, the phase differences at the intersection point of $Z_{close0}(s)$ and $Z_{Load}(s)$ vary as well. Therefore, the impact of controller parameters on system stability is studied in this part.

When K_{Rvg} is set to 30, the bode diagram of $Z_{close0}(s)$ and $Z_{Load}(s)$ of load case VIII with the tuned K_{Pvg} is shown as Figure 12a. The proportional gain K_{Pvg} = 0.8 of the AC voltage controller guarantees that the phase difference is lower than 180°. In addition, the increasing K_{Pvg} reduces the phase differences, but its influence on system stability is quite small.

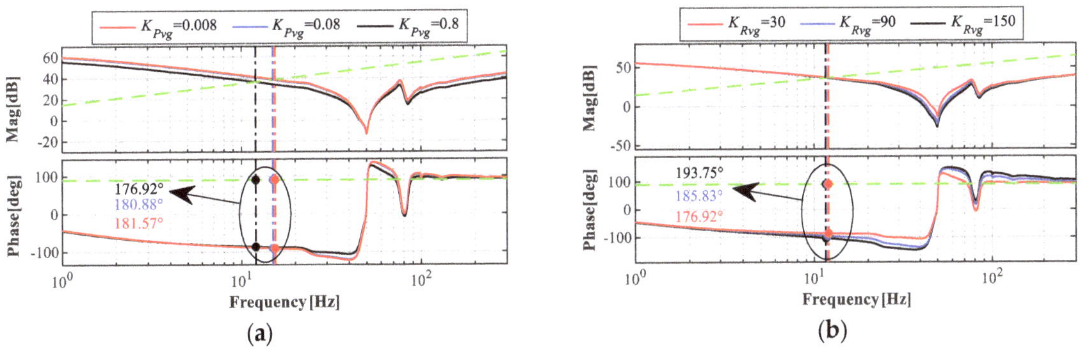

Figure 12. Bode diagrams of $Z_{close0}(s)$ and $Z_{Load}(s)$ of load case VIII with different controller parameters (Soid lines: $Z_{close0}(s)$, dashed green lines: $Z_{Load}(s)$). (**a**) Changing K_{Pvg}; (**b**) changing K_{Rvg}.

As for the resonant gain K_{Rvg}, it significantly impacts the system's stability, as shown in Figure 12b, where K_{Pvg} is set to 0.8. By tuning the resonant gain, it can be observed that the phase differences at the intersection point of $Z_{close0}(s)$ and $Z_{Load}(s)$ are gradually reduced with the decrease of K_{Rvg}. A small K_{Rvg} = 30 can ensure the system's stability.

When the resonant gains further decrease, e.g., 0.01, as shown in Figure 13a, the phase difference reduces to 172.27° but is still close to 180°. However, in this case, the AC voltage cannot track its reference value as shown in Figure 13b.

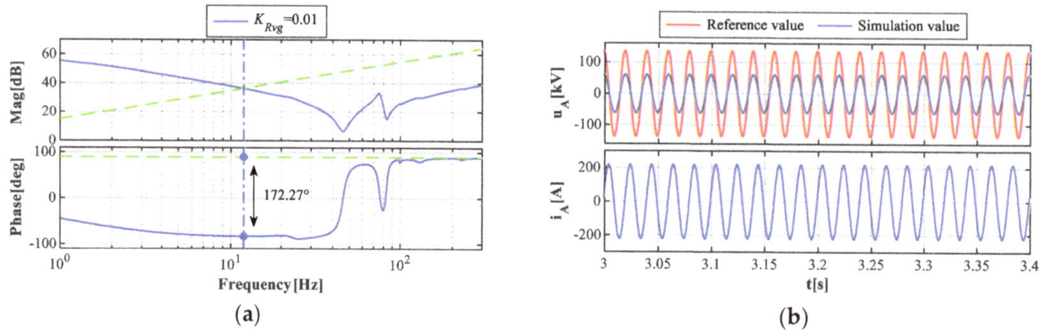

Figure 13. Bode diagrams and time-domain waveforms of load case VII (Soid lines: $Z_{close0}(s)$, dashed green lines: $Z_{Load}(s)$). (**a**) $Z_{close0}(s)$ and $Z_{Load}(s)$; (**b**) output voltage and current.

From the discussion above, it can be found that tuning the parameters of the PR controller can guarantee the system's stability. But the phase difference is relatively large and nearly impossible to further decrease. Therefore, an improved controller structure is proposed in this paper.

5.2. Proposed Controller Structure

The structure of the proposed controller is shown in Figure 14, which contains three parts:

(1) The first part is the traditional PR controller, which provides a high gain at the fundamental frequency, allowing accurate tracking of the reference signal. K_{Rvg} is the key parameter to guarantee the tracking performance.

(2) A band reject filter (BRF) is introduced, which is the second part. Based on the obtained internal DM small-signal perturbation to AC voltage transfer function, it is found that a large resonant peak appears, which affects the control performance and may threaten the system stability. The center frequency of the introduced BRF is set to suppress the resonant peak frequency at 28 Hz, and its bandwidth matches the width of the resonant peak, with a value set to 20 Hz.

(3) The third part is an additional inertia element (IE), which is added in the control loop to improve the poor stability margin of the existing PR controller. Based on the bode diagram of $Z_{close0}(s)$ and $Z_{Load}(s)$ under the worst case, the gain K_{IEvg} and cutoff angular frequency ω_{ie} of the IE are designed.

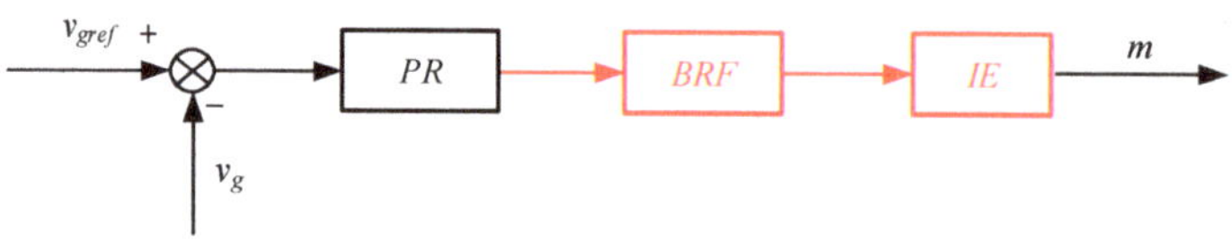

Figure 14. Block diagrams of the proposed controller structure.

The transfer function of the proposed controller can be written as

$$G_{vg}(s) = (K_{Pvg} + \frac{2K_{Rvg}\omega_c s}{s^2 + 2\omega_c s + \omega_1^2}) \cdot \frac{K_{Bvg}(s^2 + \omega_b^2)}{s^2 + 2\omega_{cb} s + \omega_b^2} \cdot \frac{K_{IEvg}}{s + \omega_{ie}} \tag{50}$$

Based on (50), the proposed controller has more control degrees of freedom, indicating that the corresponding system has better abilities in stability and control performance improvement. Figure 15 shows the flowchart of the parameter design of the proposed controller. And the corresponding parameters are given in Table 4. Figure 16 shows the bode diagram of the open-loop gain of the MMC system with the PR controller and proposed controller of the load case VIII. It can be observed that the resonant peak is suppressed, which verified the effectiveness of BRF. Moreover, the gain at the fundamental frequency remains relatively high, showing good tracking performance.

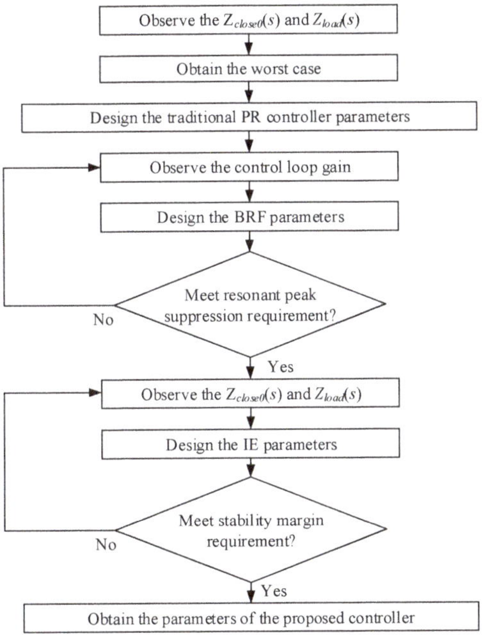

Figure 15. Flowchart of the parameter design of the proposed controller.

Table 4. Parameters of the proposed controller.

Controller Parameters	K_{Pvg}	K_{Rvg}	ω_c	K_{Bvg}	ω_b	ω_{cb}	K_{IEvg}	ω_{ie}
Values	0.8	30	π	1	56 π	40 π	200	10 π

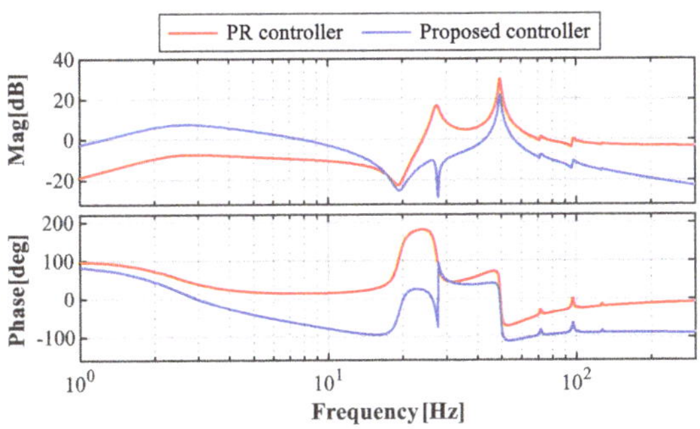

Figure 16. Bode diagram of the open-loop gain of the MMC with the traditional PR controller and the proposed controller.

Figure 17 shows the bode diagram of $Z_{close0}(s)$ and $Z_{Load}(s)$ of all load conditions with the proposed controller. It can be seen that the phase difference at the intersection point of $Z_{close0}(s)$ and $Z_{Load}(s)$ of load case VIII is equal to 114.57°. The stability margin is greatly increased, which confirms the validity and effectiveness of the proposed controller structure.

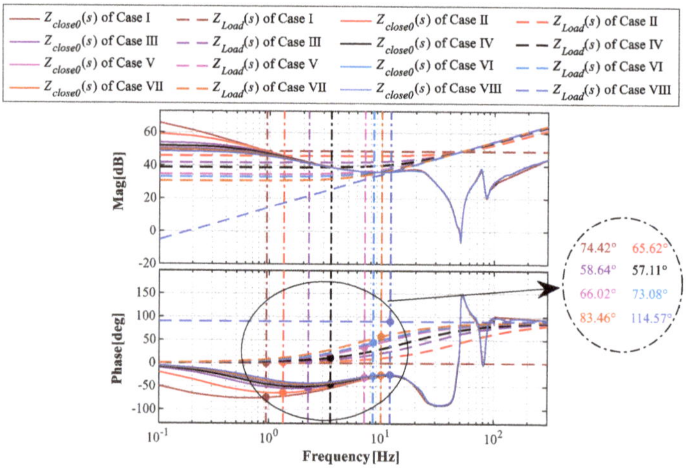

Figure 17. Bode diagrams of MMC of different load conditions with the proposed controller.

Moreover, Figure 18 shows the time-domain simulation results of the MMC with the PR controller and the proposed controller in the transient state. The AC load switches from case II to case V at t = 3.1 s. The results indicate that the MMC system reaches a stable state. However, the settling time with the proposed controller is shorter, which verifies the effectiveness of the proposed controller. The oscillation of the circulating current with the proposed controller is also smaller, as shown in Figure 18c,d. Additionally, in Figure 18e,f, the capacitor voltages also exhibit similar characteristics, further confirming the effectiveness of the proposed controller structure.

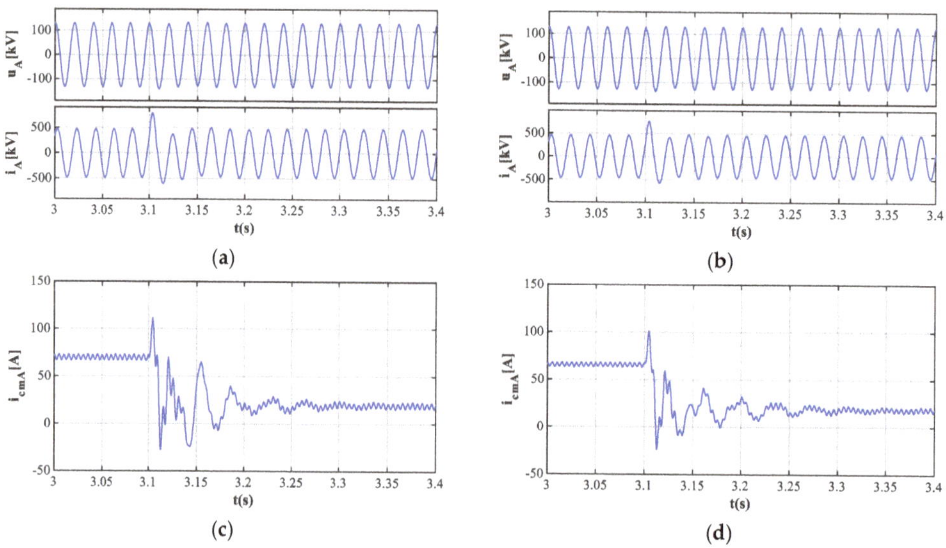

Figure 18. *Cont.*

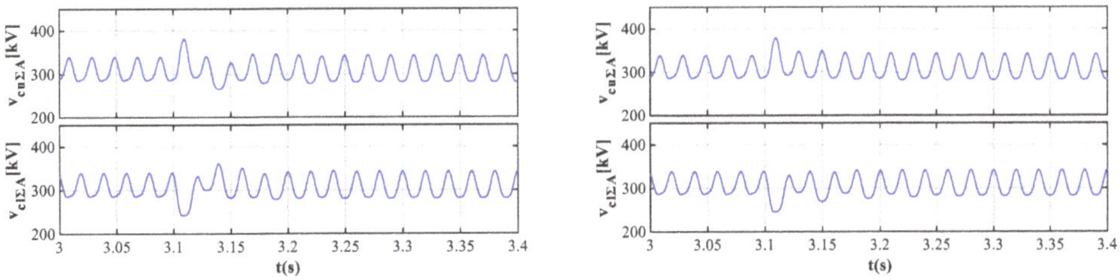

(**e**) (**f**)

Figure 18. Transient-state time-domain waveforms from case II to case V. Output voltage and current: (**a**) traditional PR controller; (**b**) proposed controller. Circulating current: (**c**) traditional PR controller; (**d**) proposed controller. Sum capacitor voltages for the upper and lower arms: (**e**) traditional PR controller; (**f**) proposed controller.

Furthermore, Figure 19 shows the corresponding time-domain simulation results, where the AC load switches from case V to case VIII at t = 3.1 s. In Figure 19a, large oscillations appear in the output voltage and current. Although the system finally reaches a stable state, the settling time is ~2.6 s, which indicates the poor control performance of the traditional PR controller. Figure 19b shows the simulation results with the proposed controller. It can be observed that the voltage and current quickly reach a stable state within only ~0.2 s, which further verifies the effectiveness of the proposed controller structure. In Figure 19c, the oscillation of the circulating current gradually decreases, with overshoot amplitude reaching ~100 A. However, for the MMC with the proposed controller, the overshoot amplitude of the circulating current is only ~50 A, as shown in Figure 19d. Additionally, Figure 19e,f indicate that the capacitor voltages have better dynamic performance with the proposed controller. In Figure 19e, the sum capacitor voltages for the upper and lower arms exhibit large overshoots, which reach ~73 kV. Figure 19f shows that the overshoot amplitudes of the sum capacitor voltages for the upper and lower arms are ~65 and ~37 kV, respectively. This further verifies the effectiveness of the proposed controller structure.

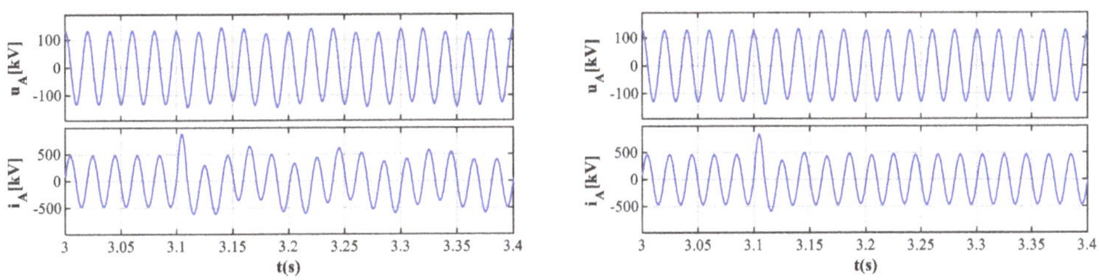

(**a**) (**b**)

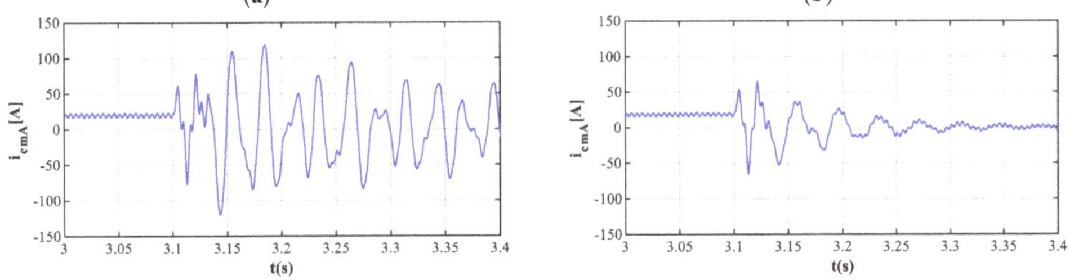

(**c**) (**d**)

Figure 19. *Cont.*

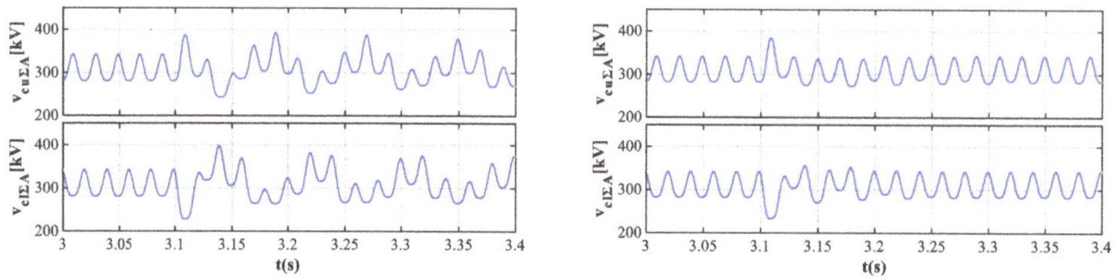

(e) (f)

Figure 19. Transient-state time-domain waveforms from case V to case VIII. Output voltage and current: (**a**) traditional PR controller; (**b**) proposed controller. Circulating current: (**c**) traditional PR controller; (**d**) proposed controller. Sum capacitor voltages for the upper and lower arms: (**e**) traditional PR controller; (**f**) proposed controller.

6. Conclusions

This paper investigates the stability variation in MMC in standalone mode. Based on the HSS, the impacts of load variation on the transfer function of internal small-signal perturbation to state variables and AC terminal impedance of the MMC are analyzed in detail. It is found that the system tends to be unstable with the increase in inductive load. Therefore, a worst-case design needs to be adopted to guarantee the overall stability. Then, the impact of traditional PR controller parameters on system stability is analyzed. It is observed that the stability margin and control performance are limited. Thus, this paper proposes an improved controller structure with an additional series compensator, which can enhance the stability and suppress the resonant peak. And the transient performance is also improved. The simulation results align well with the analysis, validating the effectiveness of the proposed controller. Based on these findings, to guarantee the stability and control performance, it is recommended that a detailed small-signal analysis of all load conditions be carried out for the MMC system.

Author Contributions: Methodology, software, and writing, X.L.; formal analysis, validation, and writing, S.S.; formal analysis, investigation, W.M.; resources, funding acquisition, Y.W. All authors have read and agreed to the published version of the manuscript.

Funding: This research was funded by the Fundamental Research Funds for the Central Universities, grant number 21CX06039A.

Data Availability Statement: Data is contained within the article.

Conflicts of Interest: The authors declare no conflicts of interest.

References

1. Liu, W.; Li, K.J.; Liu, Z.; Wang, M. A Simple and Novel Precharging Control Strategy for Modular Multilevel Converter. *IEEE Access* **2019**, *7*, 170500–170512. [CrossRef]
2. Lyu, J.; Zhang, X.; Cai, X.; Molinas, M. Harmonic State-Space Based Small-Signal Impedance Modeling of a Modular Multilevel Converter with Consideration of Internal Harmonic Dynamics. *IEEE Trans. Power Electron.* **2019**, *34*, 2134–2148. [CrossRef]
3. Perez, M.A.; Ceballos, S.; Konstantinou, G.; Pou, J.; Aguilera, R.P. Modular Multilevel Converters: Recent Achievements and Challenges. *IEEE Open J. Ind. Electron. Soc.* **2021**, *2*, 224–239. [CrossRef]
4. Lyu, J.; Yin, J.; Zhu, H.; Cai, X. Impedance Modeling and Stability Analysis of Energy Controlled Modular Multilevel Converter. *IEEE Trans. Power Deliv.* **2023**, *38*, 1868–1881. [CrossRef]
5. Zou, C.; Rao, H.; Xu, S.; Li, Y.; Li, W.; Chen, J.; Zhao, X.; Yang, Y.; Lei, B. Analysis of Resonance Between a VSC-HVDC Converter and the AC Grid. *IEEE Trans. Power Electron.* **2018**, *33*, 10157–10168. [CrossRef]
6. Guo, C.; Yang, J.; Zhao, C. Investigation of Small-Signal Dynamics of Modular Multilevel Converter Under Unbalanced Grid Conditions. *IEEE Trans. Ind. Electron.* **2019**, *66*, 2269–2279. [CrossRef]
7. Zhu, S.; Liu, K.; Qin, L.; Ran, X.; Li, Y.; Huai, Q.; Liao, X.; Zhang, J. Reduced-Order Dynamic Model of Modular Multilevel Converter in Long Time Scale and Its Application in Power System Low-Frequency Oscillation Analysis. *IEEE Trans. Power Deliv.* **2019**, *34*, 2110–2122. [CrossRef]

8. Sakinci, Ö.C.; Beerten, J. Generalized Dynamic Phasor Modeling of the MMC for Small-Signal Stability Analysis. *IEEE Trans. Power Deliv.* **2019**, *34*, 991–1000. [CrossRef]
9. Wu, H.; Wang, X.; Kocewiak, Ł.H. Impedance-Based Stability Analysis of Voltage-Controlled MMCs Feeding Linear AC Systems. *IEEE J. Emerg. Sel. Top. Power Electron.* **2020**, *8*, 4060–4074. [CrossRef]
10. Xu, Z.; Li, B.; Han, L.; Hu, J.; Wang, S.; Zhang, S.; Xu, D. A Complete HSS-Based Impedance Model of MMC Considering Grid Impedance Coupling. *IEEE Trans. Power Electron.* **2020**, *35*, 12929–12948. [CrossRef]
11. Zhu, Y.; Pou, J.; Konstantinou, G. Impedance Shaping Effects of Circulating Current Controllers in Modular Multilevel Converters. In Proceedings of the 2022 International Power Electronics Conference (IPEC-Himeji 2022-ECCE Asia), Himeji, Japan, 15–19 May 2022; pp. 555–561.
12. Zhang, Y.; Chen, X.; Sun, J. Sequence Impedance Modeling and Analysis of MMC in Single-Star Configuration. *IEEE Trans. Power Electron.* **2020**, *35*, 334–346. [CrossRef]
13. Huang, T.; Yang, F.; Zhang, D.; Chen, X. High-Frequency Stability Analysis and Impedance Optimization for an MMC-HVDC Integrated System Considering Delay Effects. *IEEE J. Emerg. Sel. Top. Circuits Syst.* **2022**, *12*, 59–72. [CrossRef]
14. Wang, D.; Zhang, K. Small-Signal Stability Analysis of MMC-HVDC System Based on Hybrid Passivity. *IEEE Trans. Power Deliv.* **2024**, *39*, 29–41. [CrossRef]
15. Wang, J.; Wang, P. Decoupled Power Control for Direct-Modulation-Based Modular Multilevel Converter with Improved Stability. *IEEE Trans. Ind. Electron.* **2019**, *66*, 5264–5274. [CrossRef]
16. Xiong, X.; Yang, Y.; Wu, C.; Zhao, C.; Blaabjerg, F. Improving the Stability of Standalone MMCs by Shaping the AC Side Impedance Using Insertion Index Compensation. *IEEE J. Emerg. Sel. Top. Circuits Syst.* **2022**, *12*, 81–89. [CrossRef]
17. Gao, L.; Zhang, H.; Xu, J. An Improved Virtual Impedance Control Based Oscillation Suppression Method for MMC DC System. *CPSS Trans. Power Electron. Appl.* **2022**, *7*, 347–358. [CrossRef]
18. Huang, P.; Vanfretti, L. Multi-Tuned Narrowband Damping for Suppressing MMC High-Frequency Oscillations. *IEEE Trans. Power Deliv.* **2023**, *38*, 3804–3819. [CrossRef]
19. Guo, C.; Xu, L.; Yang, S.; Jiang, W. A Supplementary Damping Control for MMC-HVDC System to Mitigate the Low-Frequency Oscillation Under Low Inertia Condition. *IEEE Trans. Power Deliv.* **2023**, *38*, 287–298. [CrossRef]
20. Liu, W.; Wang, R.; Kerekes, T.; Dragičević, T.; Teodorescu, R. Modulation-Compensated Power Synchronization Control for Modular Multilevel Converters with Enhanced Performance Under Varying SCR. *IEEE J. Emerg. Sel. Top. Power Electron.* **2023**, *11*, 4714–4725. [CrossRef]
21. Zhu, Y.; Jiang, S.; Pou, J.; Konstantinou, G. Virtual Arm Impedance Emulation and Stability Improvement in Modular Multilevel Converters. *IEEE Trans. Power Electron.* **2024**, *39*, 3931–3942. [CrossRef]
22. Liu, Y.; Huang, M.; Tse, C.K.; Iu, H.H.C.; Yan, Z.; Zha, X. Stability and Multiconstraint Operating Region of Grid-Connected Modular Multilevel Converter Under Grid Phase Disturbance. *IEEE Trans. Power Electron.* **2021**, *36*, 12551–12564. [CrossRef]
23. Liu, Y.; Huang, M.; Zha, X.; Iu, H.H.C. Transient current estimation of MMC-HVDC systems under a grid-converter phase difference. *CSEE J. Power Energy Syst.* **2021**, *7*, 730–743.
24. Lyu, J.; Cai, X.; Molinas, M. Frequency Domain Stability Analysis of MMC-Based HVdc for Wind Farm Integration. *IEEE J. Emerg. Sel. Top. Power Electron.* **2016**, *4*, 141–151. [CrossRef]
25. Lyu, J.; Cai, X.; Molinas, M. Optimal Design of Controller Parameters for Improving the Stability of MMC-HVDC for Wind Farm Integration. *IEEE J. Emerg. Sel. Top. Power Electron.* **2018**, *6*, 40–53. [CrossRef]
26. Isik, S.; Alharbi, M.; Bhattacharya, S. An Optimized Circulating Current Control Method Based on PR and PI Controller for MMC Applications. *IEEE Trans. Ind. Appl.* **2021**, *57*, 5074–5085. [CrossRef]
27. Kundur, P. *Power System Stability and Control*; McGraw-Hill: New York, NY, USA, 1994.
28. Meng, X.; Han, J.; Pfannschmidt, J.; Wang, L.; Li, W.; Zhang, F.; Belanger, J. Combining Detailed Equivalent Model with Switching-Function-Based Average Value Model for Fast and Accurate Simulation of MMCs. *IEEE Trans. Energy Convers.* **2020**, *35*, 484–496. [CrossRef]
29. Sun, Y.; Teixeira, C.A.; Holmes, D.G.; McGrath, B.P.; Zhao, J. Low-Order Circulating Current Suppression of PWM-Based Modular Multilevel Converters Using DC-Link Voltage Compensation. *IEEE Trans. Power Electron.* **2018**, *33*, 210–225. [CrossRef]
30. Xu, Z.; Li, B.; Wang, S.; Zhang, S.; Xu, D. Generalized Single-Phase Harmonic State Space Modeling of the Modular Multilevel Converter with Zero-Sequence Voltage Compensation. *IEEE Trans. Ind. Electron.* **2019**, *66*, 6416–6426. [CrossRef]
31. Ji, K.; Chen, W.; Wu, X.; Pang, H.; Hu, J.; Liu, S.; Cheng, F.; Tang, G. High Frequency Stability Constraints Based MMC Controller Design Using NSGA-III Algorithm. *CSEE J. Power Energy Syst.* **2023**, *9*, 623–633.
32. Sun, J. Impedance-Based Stability Criterion for Grid-Connected Inverters. *IEEE Trans. Power Electron.* **2011**, *26*, 3075–3078. [CrossRef]
33. Middlebrook, R.D. Input filter considerations in design and application of switching regulators. In Proceedings of the PESC'76, IEEE Power Electronics Specialists, Conference, Cleveland, OH, USA, 8–10 June 1976; pp. 366–382.
34. Hall, S.R.; Wereley, N.M. Generalized Nyquist Stability Criterion for Linear Time Periodic Systems. In Proceedings of the 1990 American Control Conference, San Diego, CA, USA, 23–25 May 1990; pp. 1518–1525.

energies

Article

Simulation and Characterization of Micro-Discharge Phenomena Induced by Glitch Micro-Defects on an Insulated Pull Rod Surface

Shu Niu [1], Shuai Li [1], Jizhong Liang [1], Guodong Li [1], Fan Hu [1], Hai Zhang [1], Yujie Zhu [2,*], Xianhao Fan [2,*] and Chuanyang Li [2]

[1] State Grid Shanxi Electric Power Company, Taiyuan 030001, China; nius4446sk@163.com (S.N.); lsgfr4523@163.com (S.L.); chonggen@163.com (J.L.); lgd959278428@126.com (G.L.); hitszhufan@163.com (F.H.); brighthai@163.com (H.Z.)

[2] Department of Electrical Engineering, Tsinghua University, Beijing 100084, China; chuanyang_li@mail.tsinghua.edu.cn

* Correspondence: zyujiemail@gmail.com (Y.Z.); xianhao_fan@163.com (X.F.)

Abstract: The reliability of GIS (gas-insulated switchgear) circuit breakers significantly depends on the condition of the insulated pull rods, with micro-defects on their surface posing a potential risk for micro-discharges and breakdown incidents. Experimentally investigating these micro-discharges is challenging due to their minute nature. This study introduces a framework to examine the linkage between micro-defects and micro-discharges, coupled with numerical simulations of the micro-discharge process in insulated pull rods afflicted by surface infiltration flaws under operational conditions. Initially, samples containing micro-defects were sectioned via water jet cutting for microstructural analysis through white light interferometry. Subsequently, a two-dimensional axisymmetric model simulating positive corona discharge from a needle to a plate electrode was employed to derive the relationship between charged particle density and the electric field in SF_6 and air. Building on these observations, a micro-discharge model specific to micro-defects was developed. Comparative analysis of micro-discharge behaviors in SF_6 and air for identical defect types was conducted. This research framework elucidates the discharge dynamics of charged particles in SF_6 and air during micro-discharge events, shedding light on the mechanisms underpinning micro-discharges triggered by insulation rod defects.

Keywords: gas-insulated switchgear; insulated pull rods; corona discharge; simulation analysis

1. Introduction

The insulated pull rod is a crucial insulating component in gas-insulated, metal-eclosed switchgears (GISs) characterized by its slender, elongated structure and frequent involvement in disconnection scenarios [1–5]. In operational conditions, it is subjected to significant mechanical stress and impact vibrations, necessitating superior electrical insulation and mechanical strength [2,6–9]. The resin-impregnated fiberglass (RIF) process is instrumental in achieving these properties [1,10–12]. However, inadequacies in the design of the process or its parameters can lead to micro-defects, such as white spots, burrs, and poor surface infiltration, compromising the rod's integrity [2,13,14]. The presence of these micro-defects can distort the electric field, initiating micro-discharges and potentially leading to breakdown incidents. While various methods, including X-ray, non-destructive microwave, and ultrasonic tests, have been proposed for defect detection in insulation rods, comprehensive studies on the micro-discharge phenomena induced by these defects remain scarce [2,13,14].

To advance our understanding of micro-discharge phenomena triggered by micro-defects in insulated pull rods, and given the experimental challenges in direct observation, we propose a comprehensive framework. This framework encompasses three main phases: initially, sections of the insulated pull rod harboring micro-defects are segmented via water jet cutting, followed by the examination of their surface microstructures through white light interferometry for physical characterization. Subsequently, we derive a correlation curve for charged particle density concerning the electric field within SF_6 and air environments using a two-dimensional axisymmetric model of positive corona discharge from a needle to a plate electrode. The final stage involves the formulation of a micro-discharge model predicated on the observed microstructural characteristics and the derived correlation curves. Investigating micro-discharge attributes resulting from insulation rod micro-defects not only facilitates a theoretical comprehension of insulation failures but also lays the groundwork for developing theoretical models for the online monitoring of discharge signals associated with insulation pull rod defects.

2. Characterization of the Surface Defects on Insulated Pull Rods

The efficacy of the resin-impregnated fiberglass (RIF) technique is pivotal to ensuring the integrity of insulated pull rods [10–12]. The comprehensive process entails several critical steps: Initially, fiberglass fabric is rolled into a cylindrical form and placed within a sealed mold. Subsequently, under a controlled vacuum applied at the top of the mold, liquid epoxy resin is simultaneously injected at a predetermined rate from a feeder at the bottom. The process concludes with the resin's thorough impregnation and subsequent curing phase. A misalignment in the synchronization between the vacuum and resin injection rates can introduce air into the interface between the flowing resin and the mold's interior, leading to areas of poor surface infiltration. This defect may manifest on either the inner or outer surfaces of the insulated pull rod, compromising its performance.

The predominant method for inspecting the quality of insulated pull rods is artificial visual inspection. This method, however, faces significant limitations in accurately distinguishing between inner poor infiltration defects and white spot defects. As shown in Figure 1a, both defect types manifest as irregular, gray shadows in optical transmission imaging, differing markedly from the appearance of unaffected areas. Consequently, these defects are collectively categorized as white spot defects due to their visual similarity. Notably, a distinguishing factor is the generally higher grayscale intensity observed in the light transmission images of inner poor infiltration defects compared to white spot defects. Additionally, areas of poor outer surface infiltration are identifiable by a distinct rough texture upon tactile examination.

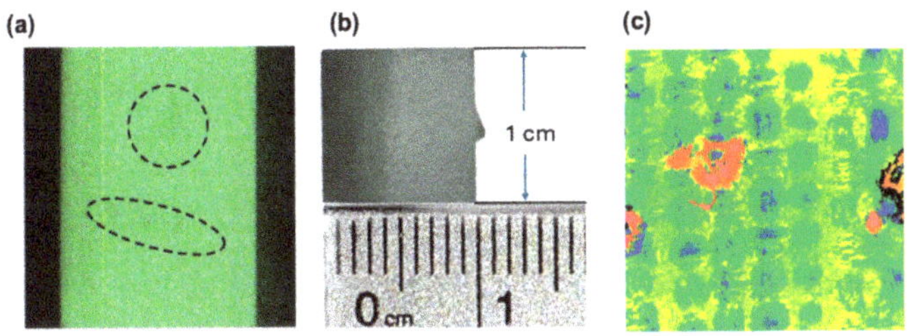

Figure 1. (**a**) Defects observed in the insulating pull rod through photonic inspection. The dashed scircles highlight areas with varying levels of light transmittance. They illustrate the location and size distribution of defects within the rod. (**b**) Dimensions of the sample sectioned for analysis. (**c**) Surface of the sample observed via white light interferometry.

With the cooperation of the factory, we identified that the minor defects in the rods mainly include five categories: internal microcracks, internal bubbles, internal impurities, surface scratches, and surface glitches. Surface scratches primarily result from damage caused by sharp tools during transportation and assembly. Glitches mainly arise from residues of the demolding material on the rod's surface during the demolding process. The main difference between the two is that scratches are more often inward depressions, while burrs protrude outward. Our study is mainly focused on surface glitch defects.

To elucidate the nature of poor infiltration defects and facilitate accurate three-dimen sional reconstruction, a 10 mm × 10 mm × 2 mm square sample from the defective area of an insulated rod was extracted using water jet cutting. This preparation enabled the differentiation of internal surface infiltration defects from white spot defects. Figure 1b showcases a section of the sample with poor surface infiltration, characterized by a macroscopically rough surface texture. Notably, these internal defects, sealed within the rod during assembly, are exposed to air under operational conditions, whereas external defects are in contact with SF_6 gas.

Microscopic examination through white light interferometry revealed that these defects consist of densely packed protrusions (Figure 1c). In the image, the red areas represent the burrs on the surface, while the surrounding green areas indicate the smooth parts of the sample. Additionally, there are some interlaced blue sections aside from the green parts. This is because the sample is a mixture of epoxy resin and glass fibers, which possess a certain level of translucency. Therefore, during white light interferometry, some light passes through the sample, causing interference. The blue areas in the image reflect the internal glass fiber framework. However, since these areas are largely opaque, this does not affect the results of the white light interferometry. The resulting surface roughness metrics are as follows: S_a (arithmetic mean surface roughness) = 6.592 μm; S_q (root mean square surface roughness) = 10.005 μm; and S_Z (peak-to-valley height) = 159.921 μm. These measurements provide a quantitative basis for the textural characterization of poor infiltration defects.

3. Analysis of Defects under Operational Conditions

Extensive literature highlights that a set of three continuity equations, representing electrons, positive ions (SF_6^+/O_2^+), and negative ions (SF_6^-/O_2^-), integrated with Poisson's equation, can be used to effectively model and elucidate the dynamics of positive corona discharge [15–17]. This study considers four charged species: electrons, positive ions, negative ions, and photoelectrons. Equations (1)–(5) delineate our mathematical framework, incorporating the drift–diffusion of these species while accounting for their generation, drift, and dissipation. The source terms within these equations include crucial processes such as ionization, electron attachment, recombination, and photoelectron absorption. Poisson's equation facilitates the computation of electric field distribution. Additionally, the Helmholtz equation is employed to depict the light absorption curve, approximated as a sum of series, providing a fit to the empirical data.

$$\frac{\partial n_e}{\partial t} + \nabla \cdot \left(-\mu_e \vec{E} n_e - D_e \nabla n_e \right) = S_{ph} + \alpha n_e \left| \mu_e \vec{E} \right| - \eta n_e \left| \mu_e \vec{E} \right| - k_{ep} n_e N_p \quad (1)$$

$$\frac{\partial n_p}{\partial t} + \nabla \cdot \left(\mu_p \vec{E} n_p - D_p \nabla n_p \right) = S_{ph} + \alpha n_e \left| \mu_e \vec{E} \right| - k_{ep} n_e n_p - k_{np} n_n n_p \quad (2)$$

$$\frac{\partial n_n}{\partial t} + \nabla \cdot \left(-\mu_n \vec{E} n_n - D_n \nabla n_n \right) = \eta n_e \left| \mu_e \vec{E} \right| - k_{np} n_n n_p \quad (3)$$

$$\nabla^2 V = -\frac{e(n_p - n_e - n_n)}{\varepsilon_0} \quad (4)$$

$$\begin{cases} \nabla^2 S_{ph}^j - \left(\lambda_j P_{SF_6}\right)^2 S_{ph}^j = -A_j P_{SF_6}^2 I_{ph} \\ S_{ph} = \sum S_{ph}^j \end{cases}, \ (j = 1, 2, 3) \quad (5)$$

where t (s) is time; n_e, n_n, n_p, μ_e, μ_n, μ_p, D_e, D_n, D_p are the different densities (m^{-3}), mobilities ($m^2V^{-1}s^{-1}$), and diffusion coefficients (m^2s^{-1}) of electrons, positive ions, and negative ions; and α (m^{-1}) and η (m^{-1}) are the ionization and attachment coefficients, respectively. k_{ep} (m^3s^{-1}) and k_{np} (m^3s^{-1}) represent the recombination coefficient of positive ions with electrons and the recombination of positive ions with negative ions. V, e, and ε_0 denote the applied voltage (V), the electron charge (C), and the vacuum permittivity ($F \cdot m^{-1}$), respectively. $\vec{E}$ ($V \cdot m^{-1}$) represents the electric field intensity vector. A_j ($cm^{-2}Torr^{-2}$) and λ_j ($cm^{-1}Torr^{-1}$) represent the fitting parameter, and the photoionization term in equation j is represented as S_{ph} ($mol \cdot m^{-3}s^{-1}$). P_{SF6} (Torr) represents the pressure of SF_6, I_{ph} ($mol \cdot m^{-3}s^{-1}$) represents the photoionization radiation coefficient.

Morrow has presented a detailed collection of reliable data on the basic transport properties of SF_6 and the approximate analytical interpolation of the coefficients [18,19]. Tables 1 and 2 show the parameters used in the simulation under air and SF_6.

Table 1. Parameters used in the air condition simulation.

Parameter	Air
Ionization coefficient α (m^{-1})	$3.5 \times 10^5 \exp\left(-1.65 \times 10^7 / \left\lvert\vec{E}\right\rvert\right)$
Attachment coefficient η (m^{-1})	$1.5 \times 10^3 \exp\left(-2.5 \times 10^6 / \left\lvert\vec{E}\right\rvert\right)$
Mobility of electrons μ_e ($m^2V^{-1}s^{-1}$)	$0.606 \times \left\lvert\vec{E}\right\rvert^{-0.25}$
Mobility of positive ions μ_p ($m^2V^{-1}s^{-1}$)	2.43×10^{-4}
Mobility of negative ions μ_n ($m^2V^{-1}s^{-1}$)	2.7×10^{-4}
Diffusion coefficient of D_e (m^2s^{-1})	0.18
Diffusion coefficient of positive ions D_p (m^2s^{-1})	2.8×10^{-6}
Diffusion coefficient of negative ions D_n (m^2s^{-1})	4.3×10^{-6}
Coefficient of recombination of positive ions and electrons k_{ep} (m^3s^{-1})	2×10^{-13}
Coefficient of recombination of positive ions and negative ions k_{np} (m^3s^{-1})	2×10^{-13}

Table 2. Parameters used in the SF_6 condition simulation.

Parameter	SF_6 Value
Ionization coefficient α (m^{-1})	$\begin{cases} 6 \times 10^{-5}, \frac{E}{N} < 1.2 \times 10^{-19}\ V \cdot m^2 \\ 1.216 \times 10^{-5} \ln(\frac{E}{N}) + 5.89 \times 10^{-4}, 1.2 \times 10^{-19} \leq \frac{E}{N} < 3.5 \times 10^{-19}\ V \cdot m^2 \\ -1.897 \times 10^{-5} \ln(\frac{E}{N}) - 7.346 \times 10^{-4}, \frac{E}{N} \geq 3.5 \times 10^{-19}\ V \cdot m^2 \end{cases}$
Attachment coefficient η (m^{-1})	$\begin{cases} 2.0463 \times 10^{-20} - 0.25379(\frac{E}{N}) + 1.4705 \times 10^{18} (\frac{E}{N})^{2.985}, \frac{E}{N} < 5 \times 10^{-20}\ V \cdot m^2 \\ -3.0078 \times 10^{36} (\frac{E}{N})^3, 5 \times 10^{-20} \leq \frac{E}{N} < 2 \times 10^{-19}\ V \cdot m^2 \\ 7 \times 10^{-21} \exp(-2.25 \times 10^{18} \frac{E}{N}), \frac{E}{N} \geq 2 \times 10^{-19}\ V \cdot m^2 \end{cases}$
Mobility of electrons μ_e ($m^2V^{-1}s^{-1}$)	$1.027 \times 10^{19} (\frac{E}{N})^{-0.2576}\ m^{-1}V^{-1}s^{-1}$
Mobility of positive ions μ_p ($m^2V^{-1}s^{-1}$)	$\begin{cases} 6 \times 10^{-5}, \frac{E}{N} < 1.2 \times 10^{-19}\ V \cdot m^2 \\ 1.216 \times 10^{-5} \ln(\frac{E}{N}) + 5.89 \times 10^{-4}, 1.2 \times 10^{-19} \leq \frac{E}{N} < 3.5 \times 10^{-19}\ V \cdot m^2 \\ -1.897 \times 10^{-5} \ln(\frac{E}{N}) - 7.346 \times 10^{-4}, \frac{E}{N} \geq 3.5 \times 10^{-19}\ V \cdot m^2 \end{cases}$
Mobility of negative ions μ_n ($m^2V^{-1}s^{-1}$)	$1.69 \times 10^{32} (\frac{E}{N})^2 + 5.3 \times 10^{-5}\ m^{-1}V^{-1}s^{-1}$
Diffusion coefficient of D_e (m^2s^{-1})	$8.8823 \times 10^{28} (\frac{E}{N})^{0.2424}\ m^{-1}s^{-1}$
Diffusion coefficient of positive/negative ions $D_{p,n}$ (m^2s^{-1})	$\begin{cases} 2 \times 10^{-13} p^{0.6336}, 1 < P \leq 39\ kPa \\ 2.28 \times 10^{-11} P^{-0.6559}, 39 < P \leq 270\ kPa \\ 6.87 \times 10^{-10} P^{-1.279}, 270 < P \leq 2000\ kPa \end{cases}$
Coefficient of recombination of positive ions and electrons/negative ions $k_{ep,np}$ (m^3s^{-1})	$\frac{D}{\mu} = \frac{k_T}{39.6}\ V$

This study employs a 2D axisymmetric needle-to-plate electrode model to investigate positive corona discharge phenomena in SF_6 and air environments. Both gases are simulated under specific conditions: SF_6 at a pressure of 0.4 MPa and air at 0.1 MPa, filling the gap between the needle and plate electrodes, with the corresponding temperature being room temperature (300 K). The geometric configuration for both gas settings is identical, featuring a needle tip curvature of 0.1 mm and a 100 mm gap between the needle tip and the plate electrode. A positive voltage ranging from 1 to 10 kV is applied to catalyze the positive corona discharge in both gases.

The study quantitatively captures the dynamics of positive corona discharge by plotting the variation in the total numbers of SF_6 electrons, negative ions, and positive ions across a discharge cycle spanning 100 ns to 1000 ns, illustrated in Figure 2. Similarly, for air, the total counts of electrons, negative ions, and positive ions over a discharge cycle ranging from 500 ns to 5000 ns are documented, as shown in Figure 3. The data points reflect the results of the simulation. The lines connecting these data points are fitting curves based on quadratic interpolation of the data points. These are used to predict missing data values. It can be observed that the positive corona discharge intensity of SF_6 and air gradually increases with the voltage applied to the needle electrode.

As the discharge progresses to 100 ns for SF_6 and 500 ns for air, the rate of change in the total particle count of both gases stabilizes, eventually reaching a steady concentration. This stability allows for the calculation of particle density by dividing the steady-state total particle count by the volume of the integral domain of each gas environment. The flux of particles generated near the glitch defect is then calculated for the next step analysis. Here, our statistical method involves using the needle tip as the center to define a circle with a radius of 2 cm as the statistical area. The electric field strength within this area is considered the true electric field strength. We measure the electron density, positive ion density, and negative ion density within this area over a time interval of 0–5000 ns as the particle yield.

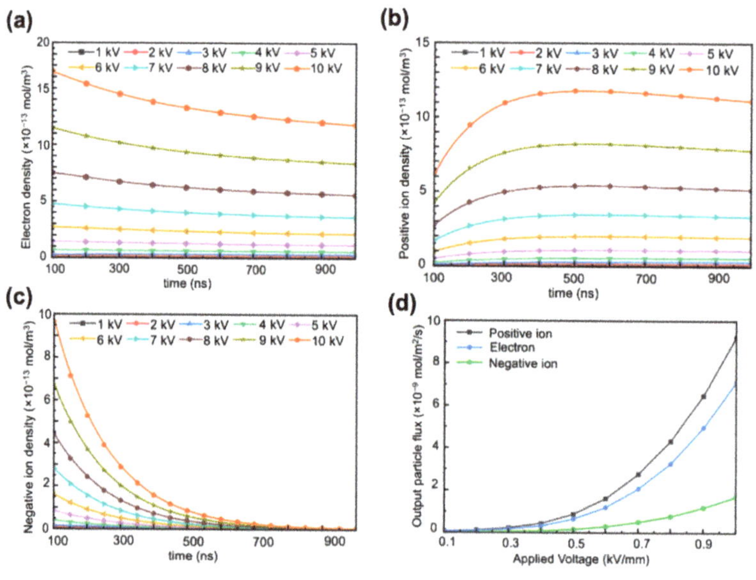

Figure 2. Through high-throughput simulation, the average densities of (**a**) electrons, (**b**) positrons, and (**c**) negative ions within different electric fields in an SF_6 environment were obtained, and (**d**) the flux of particles generated in the defect area were quantified.

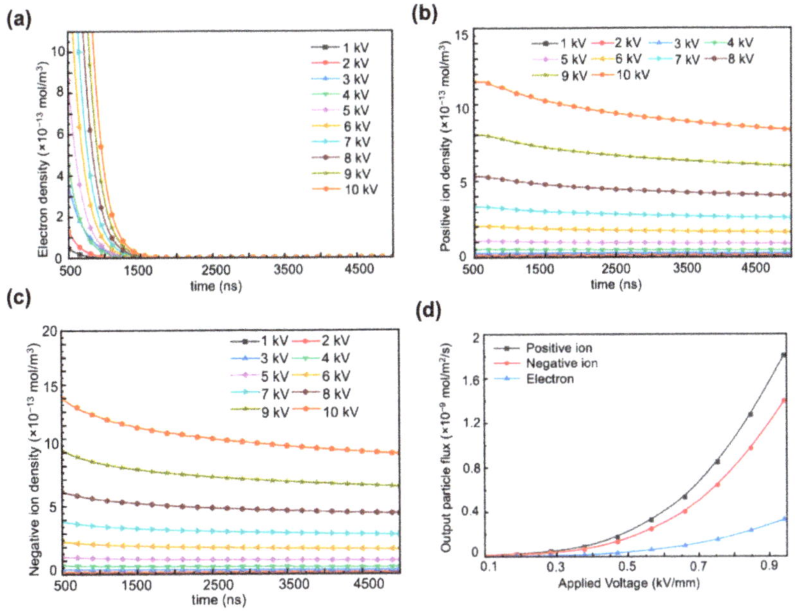

Figure 3. High-throughput simulation yielded the average densities of (**a**) electrons, (**b**) positrons, and (**c**) negative ions within different electric fields in a standard atmospheric pressure air environment, along with (**d**) the quantified flux of particles generated in the defect area.

4. Micro-Discharge Analysis in Air and SF_6 Environments under Operational Conditions

We employed the charge generation rates near the detected defects to simulate the pulling rod under operational conditions. The simulation structure is depicted in Figure 4. Initially, the discharge process within SF_6 was simulated. To further investigate the dynamics and distribution of charged particles during the micro-discharge development process, an analysis was conducted at four critical time points: t = 25 ns, t = 100 ns, t = 200 ns, and t = 500 ns.

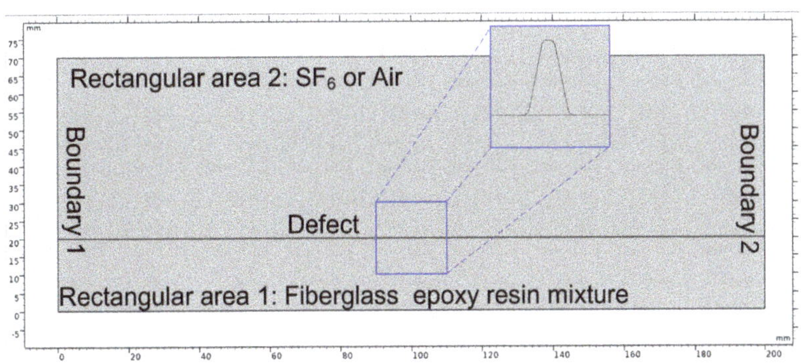

Figure 4. Schematic diagram of the simulation area.

In the initial stage (up to 100 ns), the charge distribution concentrated near the glitch defect, allowing only a few electrons to gain enough kinetic energy to initiate ionization and attachment reactions with neutral molecules. Electrons, positive ions, and negative ions were distributed evenly around the defect, as depicted in Figure 5a,e,i.

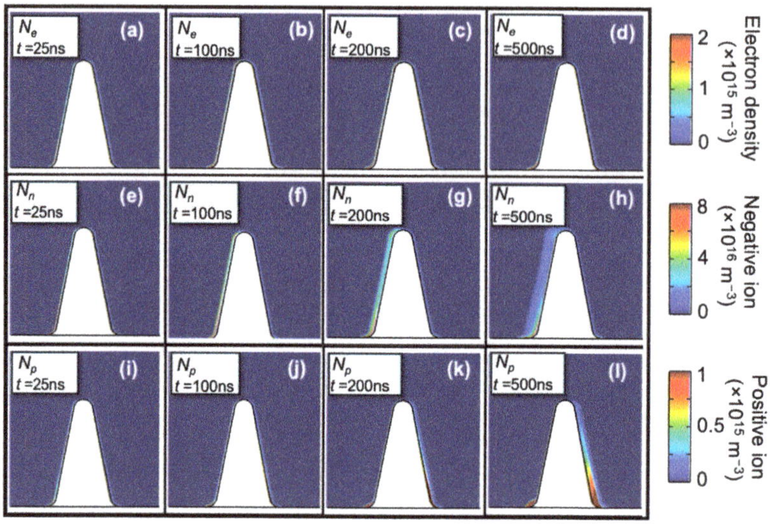

Figure 5. In an SF6 environment, the process of discharge induced by a glitch defect on the rod surface. From 25 ns to 200 ns, the evolution of (**a–d**) electrons, (**e–h**) negative ions, and (**i–l**) positive ions near the glitch defect.

As the micro-discharge developed, electrons predominantly migrated to the left side of the defect, counter to the electric field's direction, indicating the progression into later stages. SF_6's strong electronegativity plays a crucial role here, significantly enhancing electron adsorption and influencing electron density distribution throughout the micro-discharge process. Electron density peaks were observed at the defect's left corner and right wall, creating a distinct electron cloud layer. Despite moving against the initial electric field, electrons accumulate at these points. Some electrons lose their kinetic energy upon colliding with the defect's right wall, adhering to the surface and increasing electron density in this area.

Conversely, the left wall of the defect, where electron movement originates, witnesses interactions between electrons and neutral particles, producing additional positive ions and electrons. This leads to a high concentration of electron density at the left corner. However, the tendency of electrons to combine with neutral particles to form negative ions results in a lower electron density on this side.

Here, we did not consider the impact of moisture in the gas environment on the discharge at defect locations. However, in actual operation, the moisture in the gas environment can affect the defects in two main ways: the presence of moisture can influence discharge parameters such as the discharge ionization coefficient, affecting the actual discharge process; furthermore, the rod's surface can absorb moisture, increasing its surface conductivity, thereby facilitating easier migration of the charges produced by discharges across the surface.

During the SF_6 micro-discharge process, the density distribution of negative ions illustrates that SF^- is predominantly formed by electron adsorption in SF_6. Initially, electrons near the defect are captured, generating negative ions that subsequently move opposite to the electric field's direction, diffusing away from the left wall of the defect. This pattern suggests that negative ions are consistently produced and dispersed outward from the defect throughout the micro-discharge.

In the case of positive ions in SF_6, they are primarily created when SF_6 molecules collide with electrons, resulting in their detachment. Throughout the micro-discharge, there is a noticeable concentration of positive ions along both the left and right walls of the defect.

These ions show a tendency to diffuse slightly outward, aligning with the electric field's direction. Unlike electrons and negative ions, however, their diffusion is less extensive.

In an air environment, the intensity of micro-discharge around poorly insulated defects is higher than in SF_6, due to air's lower electronegativity. The evolution of micro-discharge is depicted at six critical times: t = 10 ns, t = 25 ns, t = 50 ns, t = 75 ns, t = 100 ns, and t = 500 ns, with simulation results illustrated in Figure 6.

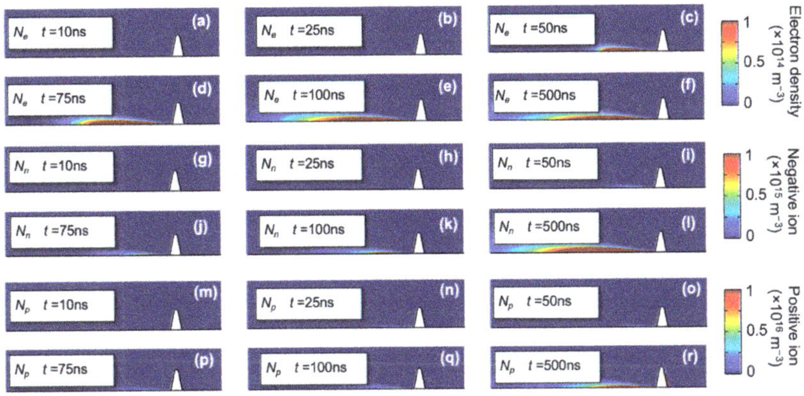

Figure 6. In an air environment, the process of discharge induced by a glitch defect on the rod surface. From 10 ns to 200 ns, the evolution of (**a**–**f**) electrons, (**g**–**l**) negative ions, and (**m**–**r**) positive ions near the glitch defect.

Initially, electron accumulation at the corner of the defect's left wall forms a thin cloud layer. Within 25 ns, there is significant electron buildup on the surface of the epoxy resin. Driven by the electric field, these electrons ionize neutral particles, leading to the generation of additional electrons and positive ions. By 50 ns, a charge cluster pattern emerges, which maintains a consistent electron cloud shape up to 500 ns. By the end of the period, there is a slight diffusion of low-density electrons observed at 500 ns.

In an air environment, negative ions are primarily produced through the adsorption of electrons onto N_2 molecules. Initially, these negative ions accumulate around the defect and migrate against the electric field, showing slower movement compared to electrons, which leads to their initial retention. By 50 ns, there is a significant accumulation of negative ions, which begins to dissipate outward by 500 ns.

The evolution of positive ions during the micro-discharge process is influenced by two main factors. Firstly, there is an accumulation of charge-like ions generated by minor ionization near the defect, distributed to the right of the defect, opposite the electric field. Secondly, new positive charges are produced by the collision ionization of surface electrons as they propagate on the left side of the defect, consistently accumulating on the surface of the insulating rod and forming a distinct charge pattern.

This study on micro-discharge phenomena caused by poor insulation defects in different environments of SF_6 and air has enhanced the understanding of the insulating performance of gas-insulated switchgear equipment. This highlights how the insulating characteristics of SF_6 and air directly influence the intensity and properties of micro-discharges triggered by such defects. Specifically, SF_6 exhibits superior insulating capabilities compared to air, effectively reducing the extent of micro-discharges under identical operational conditions. The contrasting operational environments of the inner and outer walls of the insulating rod further illustrate this point: the outer wall is exposed to SF_6 gas at a higher pressure of 4 atm, while the inner wall is surrounded by air at a lower pressure of 1 atm. This difference in gas type and pressure not only underscores the superior insulating properties of SF_6 over air but also amplifies the variation in discharge intensity between defects located on the inner and outer surfaces.

Additionally, the study of micro-discharge phenomena also reveals how insulation defects behave under different pressure and electric field conditions. This is particularly important in transformers and other high-voltage equipment, where environmental conditions significantly affect micro-discharges. As pressure increases, the insulating properties of SF6 improve, reducing both the frequency and intensity of micro-discharges. This effect has been confirmed experimentally; when the pressure of SF6 gas is increased from 1 to 4 atmospheres, micro-discharges from the same defect are noticeably weaker.

In researching micro-discharges, technologies like high-speed cameras and electron microscopes allow for a detailed observation and analysis of the dynamic processes and factors influencing micro-discharges. These precise observation methods enable researchers to accurately track the movements of electrons and ions, leading to a better understanding of how micro-discharges develop and their specific impacts on insulation performance.

This deep understanding of micro-discharge behavior provides crucial information for improving and optimizing the design of gas-insulated switchgears. By comparing how micro-discharges occur in different mediums like SF_6 and air, engineers can design safer and more efficient systems to meet the complex demands of electrical power systems. Such research not only advances high-voltage insulation technology but also supports the stability and safety of electrical power systems.

5. Conclusions

This paper introduces a simulation framework for studying micro-discharge phenomena caused by defects in insulation materials. It explores defect characterization and the processes of micro-discharges through simulations that capture temporal density fluctuations of electrons, negative ions, and positive ions, enhancing our understanding of their dynamic behavior. The study details how electric field distortion near defects leads to charged particle generation, influencing micro-discharge activity. It also examines the impact of environmental conditions, such as gas composition and pressure, on micro-discharge intensity and highlights the importance of quality control in manufacturing to improve the reliability and safety of gas-insulated switchgear systems.

Author Contributions: Methodology, S.L.; Software, G.L. and H.Z.; Validation, G.L. and H.Z.; Formal analysis, S.L. and F.H.; Resources, S.N., J.L. and F.H.; Writing—original draft, S.N., Y.Z. and X.F.; Writing—review and editing, C.L.; Project administration, J.L. All authors have read and agreed to the published version of the manuscript.

Funding: This work was supported by the State Grid Corporation Headquarters Technology Project (under Grant No. 520530230017).

Data Availability Statement: The data supporting the findings of this study are available from the corresponding author upon reasonable request.

Conflicts of Interest: Authors Shu Niu, Shuai Li, Jizhong Liang, Guodong Li, Fan Hu and Hai Zhang were employed by the State Grid Shanxi Electric Power Company. The remaining authors declare that the research was conducted in the absence of any commercial or financial relationships that could be construed as a potential conflict of interest.

References

1. Jiang, Z.J.; Li, X.; Zhang, H.; Zhang, E.Z. Research Progress and Prospect of Condition Assessment Techniques for Oil–Paper Insulation Used in Power Systems: A Review. *Energies* **2024**, *17*, 2089. [CrossRef]
2. Fan, X.; Niu, S.; Luo, H.; Liang, J.; Liu, F.; Li, W.; Liu, W.; Gao, W.; Huang, Y.; Li, C.; et al. Photon counting technique as a potential tool in micro-defect detection of epoxy insulation pull rod in gas-insulated switchgears. *High Voltage* **2024**, *9*, 267–274. [CrossRef]
3. Fan, X.; Luo, H.H.; Liang, F.W.; Hu, J.; Liu, W.D.; Li, C.Y.; He, J.L. Photon Count Technique as a Potential Tool for Insulation Micro-Defect Detection: Principles and Primary Results. *iEnergy* **2024**, *2*, 258–263. [CrossRef]
4. Deng, Y.; Fan, X.H.; Luo, H.H.; Wang, Y.; Wu, K.Y.; Liang, F.W.; Li, C.Y. Impact of Air Gap Defects on the Electrical and Mechanical Properties of a 320 kV Direct Current Gas Insulated Transmission Line Spacer. *Energies* **2023**, *16*, 4006. [CrossRef]
5. Liang, F.W.; Luo, H.H.; Fan, X.H.; Li, X.T.; Wang, X. Review of Surface Charge Accumulation on Insulators in DC Gas-Insulated Power Transmission Lines: Measurement and Suppression Measures. *Energies* **2023**, *16*, 6027. [CrossRef]

6. Riechert, U.; Holaus, W. Ultra high-voltage gas-insulated switchgear-a technology milestone. *Eur. Trans. Electr. Power* **2012**, *22*, 60–82. [CrossRef]
7. Li, X.; Zhao, H.; Murphy, A.B. SF6-alternative gases for application in gas-insulated switchgear. *J. Phys. D Appl. Phys.* **2018**, *51*, 153001. [CrossRef]
8. Li, C.; Lin, C.; Chen, G.; Tu, Y.; Zhou, Y.; Li, Q.; Zhang, B.; He, J. Field-dependent charging phenomenon of HVDC spacers based on dominant charge behaviors. *Appl. Phys. Lett.* **2019**, *114*, 202904. [CrossRef]
9. Li, C.; Hu, J.; Lin, C.; Zhang, B.; Zhang, G.; He, J. Surface charge migration and dc surface flashover of surface-modified epoxy-based insulators. *J. Phys. D Appl. Phys.* **2017**, *50*, 065301. [CrossRef]
10. Liu, Z.; Tu, T.; Euvrard, E. Combining dry type resin impregnated fiberglass paperless transformer bushings with built-in novel insulation monitoring function. In Proceedings of the 2014 IEEE PES T&D Conference and Exposition, Chicago, IL, USA, 14–17 April 2014.
11. Borodulin, A.S.; Marycheva, A.N.; Malysheva, G.V. Simulation of impregnation kinetics of fabric fillers in the production of fiberglass articles. *Glass Phys. Chem.* **2015**, *41*, 660–664. [CrossRef]
12. Rohatgi, V.; Patel, N.; James Lee, L. Experimental investigation of flow-induced microvoids during impregnation of unidirectional stitched fiberglass mat. *Polym. Compos.* **1996**, *17*, 161–170. [CrossRef]
13. Yan, B.; Yang, W.; Wang, H.; Hou, C.; Kong, X.; Du, B. Identification and Analysis of Micro Defects in Insulation Pull Rod Based on the Ultrasonic Method. In Proceedings of the 2023 International Symposium on Electrical Insulating Materials (ISEIM), Shimane, Japan, 24–28 September 2023; pp. 336–339.
14. Luo, H.; Fan, X.; Liang, F.; Hu, J.; Yan, B.; Li, T.; He, J.; Li, C. Surface Defect Detection of GIS Aramid Insulation Pull Rod via Photon Counting. In *IEEE Transactions on Dielectrics and Electrical Insulation*; early access; IEEE: Piscataway, NJ, USA, 2024; pp. 1–7.
15. Adamiak, K.; Atten, P. Simulation of corona discharge in point–plane configuration. *J. Electrostat.* **2004**, *61*, 85–98. [CrossRef]
16. Weingärtner, E.; Kuster, F.; Wegener, K. Modeling and simulation of electrical discharge machining. *Procedia CIRP* **2012**, *2*, 74–78. [CrossRef]
17. Ming, W.; Zhang, S.; Zhang, G.; Du, J.; Ma, J.; He, W.; Cao, C.; Liu, K. Progress in modeling of electrical discharge machining process. *Int. J. Heat. Mass. Transf.* **2022**, *187*, 122563. [CrossRef]
18. Morrow, R. The theory of positive glow corona. *J. Phys. D Appl. Phys.* **1997**, *30.22*, 3099. [CrossRef]
19. Morrow, R. Theory of stepped pulses in negative corona discharges. *Phys. Rev. A* **1985**, *32*, 3821. [CrossRef] [PubMed]

energies

MDPI

Article

A Spread-Spectrum Modulation Scheme for a 3 × 6 Indirect Matrix Converter Based on a Current Ripple Model

Zhanqing Zhou [1,2,*], Lingyue Xue [1], Chen Li [2] and Qiang Geng [1]

1 Tianjin Key Laboratory of Intelligent Control of Electrical Equipment, Tiangong University, Tianjin 300387, China; xuelingyue@tiangong.edu.cn (L.X.); gengqiang@tju.edu.cn (Q.G.)
2 Zhejiang University Advanced Electrical Equipment Innovation Center, Hangzhou 311107, China; lichen_hz@zju.edu.cn
* Correspondence: zhouzhanqing@tiangong.edu.cn

Abstract: Focused on addressing harmonic suppression in multiphase indirect matrix converters (IMCs), this study explores spread-spectrum modulation technology through ripple analysis calculations. We introduce a current ripple spread-spectrum modulation (CR-SSM) method tailored for multiphase IMC systems. In this approach, a 3 × 6-phase IMC is modeled as a two-port network, and a small-signal model of the output side is established. The transfer function is utilized to analyze the two-port network in the complex frequency domain (s-plane). The time-domain expression of the output current ripple is derived in vector form. Subsequently, the distribution of the output ripple and locus are determined based on specified constraints. The carrier frequency is dynamically adjusted online according to the specified ripple locus. Compared to classical periodic PWM methods, this approach offers a broader range of frequency variations and achieves a more uniform output spectrum. Furthermore, CR-SSM optimizes system efficiency and enhances spread-spectrum modulation. Experimental results demonstrate that this method effectively enhances the quality of input and output waveforms in multiphase IMC systems.

Keywords: indirect matrix converter (IMC); current ripple analysis; spread-spectrum modulation; periodic PWM; electromagnetic interference (EMI)

1. Introduction

The indirect matrix converter (IMC) represents a form of "all-silicon" AC/AC direct power converter [1]. This converter achieves efficient power conversion without requiring an energy storage link, owing to its compact structure, ability to deliver sinusoidal input–output waveforms, adjustable input power factor, low harmonic distortion, flexible phase adjustment of input current, and other noteworthy features. The IMC has proven highly effective in applications within the green energy sector [2,3].

In a three-phase system, the conventional configuration of a matrix converter (MC) is referred to as a direct matrix converter (DMC), featuring a 3 × 3 bidirectional switch array [4]. A multiphase indirect matrix converter (IMC) comprises a three-phase rectifier stage and a six-phase inverter stage, each divided into distinct structural and functional segments. As illustrated in Figure 1, the rectifier stage of the IMC resembles a current source rectifier (CSR). Conversely, the inverter stage of the IMC is designed to function as two two-level voltage source inverters (VSIs) [5]. The clamper circuit is specifically implemented for protective purposes and does not actively facilitate energy transfer. The inherent design of the IMC allows for direct energy transfer between input and output, but it also enables the transmission of harmonic distortion from one side of the converter to the other. The study by [6] enhances the understanding of IMC topology, indicating that the input filter plays a critical role in the converter system by attenuating high-frequency harmonics to prevent interference with the normal operation of other equipment at their common connection point. Additionally, it assists in smoothing out discrete input current to generate a continuous sinusoidal current waveform on the source side. One fundamental and widely used method for improving waveform quality in frequency converters is by increasing the carrier frequency of the controller. Recent advancements in wide bandgap semiconductors, such as silicon carbide (SiC), have significantly expanded the feasible carrier frequency of power converters [7–9]. Various approaches have been proposed to enhance the effectiveness of power converters using pulse-width modulation (PWM), including optimization techniques for vector selection among available vectors [10], reduction in common-mode components in modulation waves [11,12], refinement of switch sequences [13–15], and enhancement of carrier properties [16–18].

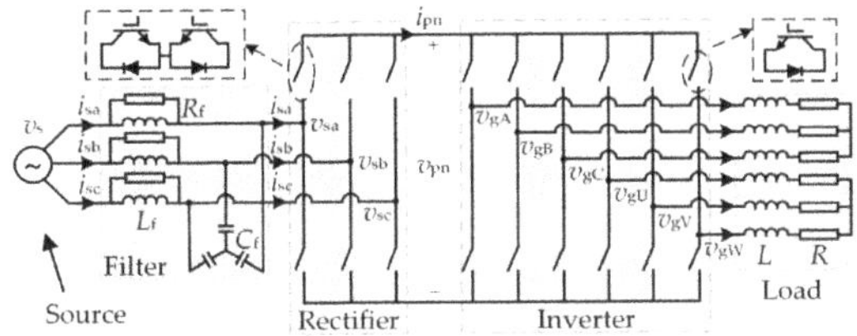

Figure 1. A 3 × 6 IMC system.

Spread-spectrum modulation (SSM) is an ideal strategy for reducing EMI in multiphase IMC systems. Common SSM strategies currently include random PWM, programmed PWM, and periodic PWM. In comparison to random PWM, periodic PWM uses a predetermined frequency density function to dynamically change the carrier frequency, achieving a fixed EMI spectrum [19]. By varying the carrier frequency, periodic PWM introduces additional modulation flexibility. Typical carrier waveforms encompass sinusoidal, exponential, and triangular shapes. Dual Fourier analysis of inverter output voltages reveals that triangular wave frequency density functions offer smoother spectral performance compared to sinusoidal and exponential carriers, which exhibit distinct spectral peaks at specific frequencies [20]. Research on the variation pattern of periodic signal frequency and waveform under equal maximum frequency deviation is explored in [21], which derives an optimal segmented modulation signal for superior EMI suppression by optimizing a combination of three frequency density functions. Furthermore, reference [22] employs a variable density sawtooth carrier to implement periodic PWM, thus achieving a spread-spectrum effect while reducing inverter switch frequencies. Reference [23] explores a spread-spectrum modulation scheme by varying switch frequencies according to uniform distribution, exponential distribution, and normal distribution and derives corresponding equations. Experimental results demonstrate that the proposed spread-spectrum modulation scheme can reduce EMI by 20 dB. Reference [24] proposes an algorithm using sinusoidal-wave pulse-width modulation (PWM) to suppress conducted-mode (CM) EMI in large-scale networks. This method minimizes switch frequency overlap between converters, thereby reducing total EMI levels. The authors of [25,26] focus on using Σ-Δ modulation for voltage source converters. The proposed modulation techniques significantly reduce switching losses and reduce lower-order harmonics by varying switch frequencies. Additionally, Y. Huang utilizes a variable frequency density sawtooth carrier to achieve periodic PWM, thereby achieving spread-spectrum effects while reducing inverter switch frequencies [27]. Reference [28] presents an optimized periodic PWM scheme that minimizes motor vibration frequency response peaks by considering the motor's vibration frequency response function. Reference [29] details the impact of periodic PWM on the current harmonics of permanent magnet synchronous motors (PMSMs). Experimental evidence demonstrates that periodic PWM disperses high-frequency harmonics across wider frequency bands, thereby suppressing peak harmonics. Reference [30] proposes a spectral analysis method based on periodic PWM and dual Fourier series to reduce current harmonics in active power filters. Reference [31] introduces a novel PWM strategy based on uniform distribution PWM and experimentally verifies its effectiveness in eliminating EMI peaks, reducing conducted EMI, and minimizing switch losses. Reference [32] analyzes the suppression of harmonics and common-mode EMI using periodic PWM, validating the accuracy of theoretical analysis and the effectiveness of algorithms through experiments.

However, the widespread implementation of current-ripple-based spread-spectrum modulation (CR-SSM) in multiphase inverter motor control (IMC) systems still encounters several challenges. Firstly, variations in vector amplitudes of the rectifier and inverter stages within the IMC, influenced by the DC-link voltage, pose difficulties in establishing a consistent ripple model. Secondly, the presence of input filters can affect waveform quality at the grid side and complicate the analysis of the impact of input switching sequences and transmission ratios on the output current ripple locus. Additionally, the resonant frequency and bandwidth of filters impose limitations on the range of carrier frequency variations.

To address these challenges, this study investigates the combined impact of the rectifier and inverter stages in a 3 × 6-phase inverter motor control (IMC) system [33,34]. It involves an analysis of a small-signal model at the output stage to derive a time-domain expression for the output current ripple. Finally, based on the constraints related to ripple distribution and ripple index, the CR-SSM (current ripple spread-spectrum modulation) method is proposed. This method allows for a broader range of frequency fluctuations while ensuring safe commutation and achieving higher-quality spread-spectrum modulation. The main contributions of this paper are outlined as follows:

1. Through the establishment of a small-signal model on the output side, the locus of the output current ripple can be determined. Furthermore, dynamically adjusting the carrier frequency based on the ripple locus can effectively broaden the harmonic spectrum, improve total harmonic distortion, and enhance output modulation efficiency.
2. The CR-SSM method dynamically adjusts the carrier frequency based on the ripple locus. The CR-SSM method is based on the output current ripple, and thus the debugging process of traditional periodic PWM is eliminated. It standardizes the construction of carrier frequencies, providing clear numerical indicators and improving frequency fluctuation efficiency. Furthermore, it reduces harmonic spikes at specific frequency bands and optimizes system efficiency.

The remainder of this paper is structured as follows: In Section 2, we provide a brief overview of multiphase IMC basic modulation strategies. Section 3 analyzes the voltage error locus and establishes small-signal models for multiphase IMC outputs to derive time-domain expressions for the current ripple. Subsequently, we explore achieving improved total harmonic distortion (THD) and efficiency by implementing online adjustment of carrier frequencies along the ripple locus. Section 4 presents experimental results comparing CR-SSM and classical period PWM. Finally, Section 5 summarizes the key conclusions drawn from this study.

2. Modulation Method of Multiphase IMC

2.1. Space Vector Modulation

To enhance the utilization of input voltage and facilitate the commutation action of the switching device, we employed the indirect space vector modulation method for the multiphase IMC. Specifically, the rectifier stage adopts zero-free vector modulation based on the expected input current vector $\boldsymbol{I}_{\mathrm{i}}^{*}$, while the inverter stage utilizes a maximum four-vector SVPWM modulation according to the anticipated output voltage vector $\boldsymbol{V}_{\mathrm{o}}^{*}$, as illustrated in Figure 2. In this paper, the moment represented by the vector position in Figure 2 is presented as an example. The red color arrows represent the vector of the rectifier stage in the I sector and the inverter stage in the ① sector.

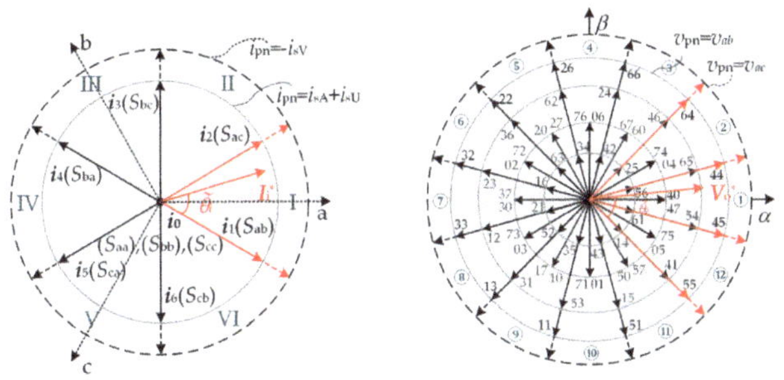

(a) (b)

Figure 2. The vectors diagram of 3 × 6 IMC drives: (**a**) rectifier stage; (**b**) α-β.

As shown in Figure 2a, the space vector plane of the rectifier is divided into six sectors, labeled as I, II, ..., VI. The symbols i_x (x = 1, 2, ..., 6) represent the rectifier active vectors, and $\widetilde{\theta}_i$ represents the angle included between $\boldsymbol{I}_i^*$ and the starting vector position of the sector.

$$\boldsymbol{i}_x = \frac{2}{\sqrt{3}} i_{pn}\, e^{j(2x-3)\pi/6} \tag{1}$$

$$\boldsymbol{I}_i^* = \frac{2}{3}(i_a + i_b\, e^{j2\pi/3} + i_c e^{j4\pi/3}) \tag{2}$$

where i_{pn} is the DC-link current, which is related to the load current and the inverter switching state. The rectifier stage is usually modulated by two effective vectors adjacent to $\boldsymbol{I}_i^*$. The rectifier stage is responsible for synthesizing v_{pn} and regulating the $\boldsymbol{I}_i^*$ phase, and the $\boldsymbol{I}_i^*$ amplitude is determined by the output current. The rectifier stage duty cycle is

$$\begin{cases} d_\mu = \sin(\frac{\pi}{3} - \widetilde{\theta}_i) / \cos(\frac{\pi}{6} - \widetilde{\theta}_i) \\ d_v = \sin(\widetilde{\theta}_i) / \cos(\frac{\pi}{6} - \widetilde{\theta}_i) \end{cases} \tag{3}$$

where d_μ is the duty cycle of the vector $\boldsymbol{i}_\mu$ (vector $\boldsymbol{i}_1$ in Figure 2); d_v is the duty cycle of the vector $\boldsymbol{i}_v$ (vector $\boldsymbol{i}_2$ in Figure 2). When the effective current vector $\boldsymbol{i}_\mu$ acts, the DC-link voltage is $\boldsymbol{v}_{ab}$ (corresponding to the input S_{ab}), and when the effective current vector $\boldsymbol{i}_v$ is acted upon, the DC-link voltage is $\boldsymbol{v}_{ac}$ (corresponding to the input S_{ac}). $\boldsymbol{V}_{im}$ is the amplitude of the three-phase sinusoidal input voltage so that the average DC-link voltage can be expressed as follows:

$$v_{pn} = v_\mu d_\mu + v_v d_v = \frac{3V_{im}}{2\cos(\pi/6 - \widetilde{\theta}_i)} \tag{4}$$

In Figure 2a, the switching state S_{ab} corresponds to the DC-link voltage v_{ab}. Figure 2b is a schematic diagram of the inverter stage output voltage space vector. The space vector plane of the inverter is divided into 12 sectors, labeled as ①, ②, ..., ⑫. The symbols v_y (y = 1, 2, ..., 12) represent the inverter active vectors; v_0 and v_7 are the zero voltage vectors; $\boldsymbol{V}_o^*$ is the desired output voltage vector; $\widetilde{\theta}_o$ represents the angle included between $\boldsymbol{V}_o^*$ and the starting vector position of the sector; and v_{pn} is the DC-link voltage.

$$\boldsymbol{v}_y = \frac{2}{3} v_{pn} e^{j(2y-1)\pi/12} \tag{5}$$

$$\boldsymbol{V}_o^* = \frac{2}{3}(v_a + v_b e^{j2\pi/3} + v_c e^{j4\pi/3} + v_u e^{j\pi/6} + v_v e^{j5\pi/6} + v_w e^{j3\pi/2}) \tag{6}$$

The IMC voltage transmission ratio q is defined as follows:

$$q = V_{\text{om}} / V_{\text{im}} \tag{7}$$

where V_{om} and V_{im} are the desired output and input voltage vector amplitudes.

According to the vector space decoupling (VSD) transformation, each voltage vector of the inverter stage can be projected to the α-β plane and the harmonic plane x-y. Each plane contains 60 valid vectors and 4 zero vectors. While synthesizing the α-β plane fundamental voltage, it is necessary for the x-y plane output voltage to be zero [34]. Thus, a minimum of four basic vectors are required to fulfill the modulation requirements, as expressed below.

$$\begin{bmatrix} d_1 \\ d_2 \\ d_3 \\ d_4 \end{bmatrix} = \begin{bmatrix} v_{1\alpha} & v_{2\alpha} & v_{3\alpha} & v_{4\alpha} \\ v_{1\beta} & v_{2\beta} & v_{3\beta} & v_{4\beta} \\ v_{1x} & v_{2x} & v_{3x} & v_{4x} \\ v_{1y} & v_{2y} & v_{3y} & v_{4y} \end{bmatrix}^{-1} \begin{bmatrix} v_\alpha^* \\ v_\beta^* \\ 0 \\ 0 \end{bmatrix} \tag{8}$$

where d_1~d_4 donate the duty cycles of the four basic vectors, namely $\boldsymbol{V}_1(55)$, $\boldsymbol{V}_2(45)$, $\boldsymbol{V}_3(44)$, and $\boldsymbol{V}_4(64)$, which are required to synthesize the output reference vector $\boldsymbol{V}_o^*$. To ensure output voltage utilization, $\boldsymbol{V}_1$, $\boldsymbol{V}_2$, $\boldsymbol{V}_3$, and $\boldsymbol{V}_4$ can be selected as four vectors that are adjacent to $\boldsymbol{V}_o^*$, $d_0 = 1 - d_1 - d_2 - d_3 - d_4$ donates the duty cycle of zero vectors. Table 1 provides the applied times of voltage vectors and the switch selection of the inverter sectors ①, ②, ... ⑫ within $T_c/2$. $\boldsymbol{V}_0$ and $\boldsymbol{V}_7$ are zero vectors.

Table 1. Voltage vectors and switch selection of the inverter sectors.

	Sec ①	**Sec ②**	**Sec ⑤**	**Sec ⑥**	**Sec ⑨**	**Sec ⑩**
V_0	00	00	00	00	00	00
V_2	45	44	26	22	13	11
V_1	55	45	66	26	33	13
V_7	77	77	77	77	77	77
V_4	64	66	32	33	51	55
V_3	44	64	22	32	11	51
V_7	77	77	77	77	77	77
	Sec ③	**Sec ④**	**Sec ⑦**	**Sec ⑧**	**Sec ⑪**	**Sec ⑫**
V_0	00	00	00	00	00	00
V_1	44	64	22	32	11	51
V_2	64	66	32	33	51	55
V_7	77	77	77	77	77	77
V_3	66	26	33	13	55	45
V_4	26	22	13	11	45	44
V_7	77	77	77	77	77	77

Given that the rectifier and inverter stages are directly cascaded, it is necessary to ensure coordination between the switching sequences of the two stages. In order to reduce switching losses, each power device is switched only once in each PWM cycle. In this paper, symmetrical triangular carrier PWM is adopted. The rectifier stage vector sequence is divided into 3 segments, and the inverter stage sequence is divided into 13 segments. The vector action sequence is shown in Figure 3. To minimize the switching action, the rectifier stage vector sequence needs to be switched from $\boldsymbol{i}_\mu$-$\boldsymbol{i}_\nu$-$\boldsymbol{i}_\mu$ to $\boldsymbol{i}_\nu$-$\boldsymbol{i}_\mu$-$\boldsymbol{i}_\nu$ when $\boldsymbol{I}_i^*$ turns from an odd sector to an even sector. Similarly, the inverter stage follows the principle of minimum switching action. T_c is a carrier cycle; $d_\mu = T_c\, d_\mu$ and $d_\nu = T_c\, d_\nu$ is the duration time of rectifier vector $\boldsymbol{i}_\mu$ and $\boldsymbol{i}_\nu$, respectively; $d_{2\mu} = T_c d_\mu d_2$ is the duration time when rectifier vector i_μ and inverter vector v_2 are implemented.

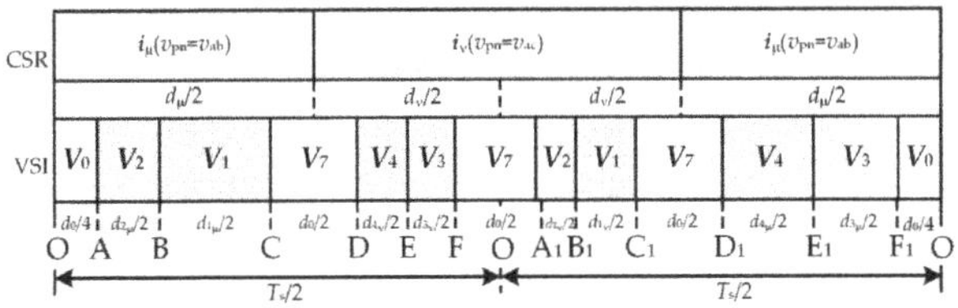

Figure 3. Switching pattern of 3 × 6 IMC.

2.2. Classical Periodic PWM

In the classical periodic PWM modulation, the carrier frequency of the multiphase IMC inverter stage is no longer constant [35–38] but varies periodically around a specific center frequency according to certain mathematical rules.

$$f_c(t) = f_{c0} \pm \Delta f_c v_m(t) \tag{9}$$

where f_{c0} represents the center frequency of the time-varying carrier frequency, while Δf_c denotes the maximum frequency deviation from the time-varying carrier frequency to the center frequency. Furthermore, v_m(t) stands for a periodic signal with an amplitude within $[-1, 1]$. Typically, sine wave, triangular wave, and exponential wave are selected as modulation signals, which can be expressed as follows:

$$m(t) = M_m \sin(2\pi f_m t) \tag{10}$$

$$\begin{cases} 0 < t < k_s\frac{T_m}{2},\ m(t) = M_m f_m \frac{2}{k_s} t \\ k_s\frac{T_m}{2} < t < (1-\frac{k_s}{2})T_m,\ m(t) = \frac{M_m}{(1-k_s)}(1-2f_m t) \\ (1-\frac{k_s}{2})T_m < t < T_m,\ m(t) = M_m \frac{2}{k_s}(f_m t - 1) \end{cases} \tag{11}$$

where k_s is the symmetry index in the range (0, 1).

$$\begin{cases} 0 < t < \frac{T_m}{4},\ m(t) = \frac{M_m}{(e^{\frac{p}{4f_m}}-1)}(e^{pt}-1) \\ \frac{T_m}{4} < t < \frac{T_m}{2},\ m(t) = \frac{M_m}{(e^{\frac{p}{4f_m}}-1)}(e^{\frac{p}{2f_m}}e^{-pt}-1) \\ \frac{T_m}{2} < t < \frac{3T_m}{4},\ m(t) = \frac{M_m}{(e^{\frac{p}{4f_m}}-1)}(1-e^{\frac{-p}{2f_m}}e^{pt}) \\ \frac{3T_m}{4} < t < T_m,\ m(t) = \frac{M_m}{(e^{\frac{p}{4f_m}}-1)}(1-e^{\frac{p}{f_m}}e^{-pt}) \end{cases} \tag{12}$$

where p is the concavity coefficient.

According to the Parseval theorem, in order to ensure constant frequency energy, the peak harmonic amplitude decreases as the bandwidth of the harmonic increases [39]. By using periodic PWM modulation, the carrier frequency is no longer fixed, so the high-frequency harmonics of multiphase IMC will not concentrate near a fixed frequency. This can greatly reduce the amplitude of high-frequency current harmonics, but there still exists some harmonic content.

3. Spread-Spectrum Modulation Based on Current Ripple

3.1. Output Voltage Error Locus

The equilibrium operating points of the vectors v_s, i_s, v_i, and i_i are the grid voltage sampling value vector $\boldsymbol{V}_s$, the average grid-side current vector $\boldsymbol{I}_s$, the average input voltage vector $\boldsymbol{V}_i$, and the input-side current expectation vector $\boldsymbol{I}_i^*$ per unit switching cycle. The disturbance signals are the ripple vectors $\widetilde{v}_s$, $\widetilde{\boldsymbol{i}}_s$, and $\widetilde{v}_i$, and the input current error vector $\widetilde{\boldsymbol{i}}_i$.

Space vector modulation utilizes fundamental voltage vectors and zero vectors to synthesize the continuously varying reference voltage $\boldsymbol{V}_o^*$. When there is a variance between the actual output voltage of the multiphase indirect matrix converter and the reference voltage, a vector model can be established to represent the error in voltage between the reference and actual output voltages [39,40].

$$\widetilde{V}_s(t) = v_o - V_o^* \tag{13}$$

where $\widetilde{V}_s(t)$ represents the voltage fluctuation quantity, v_o represents the actual output voltage vector, and $\boldsymbol{V}_o^*$ is the reference voltage. Taking $\boldsymbol{V}_o^*$ located in the first sector as an example, $\boldsymbol{V}_1$, $\boldsymbol{V}_2$, $\boldsymbol{V}_3$, and $\boldsymbol{V}_4$ are required to synthesize this vector, corresponding to inverter stages $\boldsymbol{V}_{55}$, $\boldsymbol{V}_{45}$, $\boldsymbol{V}_{44}$, and $\boldsymbol{V}_{64}$.

Taking $\boldsymbol{V}_o^*$ located in the first sector as an example, $\boldsymbol{V}_1$, $\boldsymbol{V}_2$, $\boldsymbol{V}_3$, and $\boldsymbol{V}_4$ required to synthesize this vector correspond to the inverter levels $\boldsymbol{V}_{55}$, $\boldsymbol{V}_{45}$, $\boldsymbol{V}_{44}$, and $\boldsymbol{V}_{64}$, and the magnetic chain fluctuations will be presented as five dimensions in the unit switching period, which are $v_{1\mu}$, $v_{1\nu}$, $v_{2\mu}$, $v_{2\nu}$, $v_{3\mu}$, $v_{3\nu}$, and $v_{4\mu}$, $v_{4\nu}$, along the $\boldsymbol{V}_1$~$\boldsymbol{V}_4$. As shown in the PWM vector sequence, within one switching cycle, the actual input current vector is composed of five instantaneous vectors: the direction of $\boldsymbol{i}_{\mu\alpha}$, $\boldsymbol{i}_{\mu\beta}$ along the direction of $\boldsymbol{i}_1$, and $\boldsymbol{i}_{\nu\alpha}$, $\boldsymbol{i}_{\nu\beta}$ along the direction of i_2, as well as the zero vector $\boldsymbol{i}_0$. Correspondingly, there are five types of error current vectors, as depicted in Figure 4a.

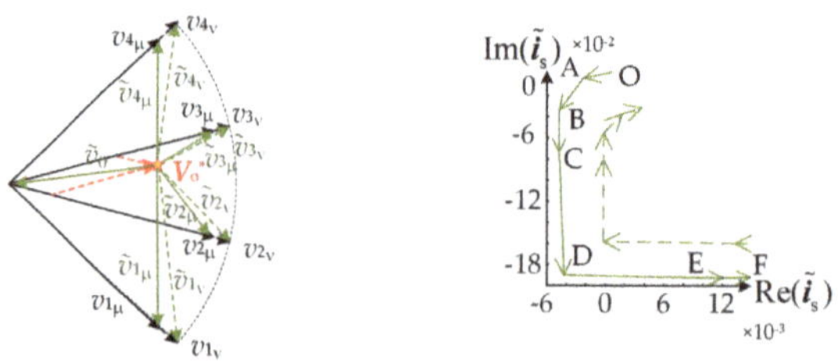

Figure 4. Error vectors and their locus in the IMC output side in Section ①: (**a**) output voltage–error vectors. (**b**) output current ripple vectors.

From this, the locus of the output current ripple vector is obtained, as shown in Figure 4b. At the beginning of the carrier cycle, $t = t_0$, v_o coincides with $\boldsymbol{V}_o^*$, and the voltage ripple vector corresponds to point O in Figure 4b; during the time period of t_0~t_7, the locus departs from point O and forms the polygon OABCDEFO along the direction of the arrow; from the vector sequence layout in Figure 2, it can be seen that the vector locus during the time period of t_0~t_7 is the same as that during the time period of t_7~t_{14}. It needs to be noted that the ratio of unit length on the Re-Im axis is set to 1:10, which is due to the fact that it is a qualitative rather than a quantitative calculation of ripples and does not mislead the understanding of its properties.

3.2. Small-Signal Model

Based on the principle of high-frequency synthesis, the space vector method employs a finite number of effective and zero vectors to continuously synthesize desired vectors with varying values. The discrepancy between the actual and desired vectors results in the input and output of the IMC containing numerous high-frequency voltage harmonics, ultimately contributing to the formation of high-frequency ripple components. It should be noted that I_i^* is usually synthesized by the two boundary current vectors in the rectifier stage. In Figure 3, one section is the action time of the boundary current vector i_μ, and the other sector is the action time of the boundary current vector i_v. Thus, the chopper DC-link voltage of the multiphase IMC will be in a two-level fluctuation state along the three-phase sine envelope. This implies that the output current ripple of the IMC is not solely determined by the switching sequence of the inverter stage but is also affected by the switching sequence of the rectifier stage. In order to mitigate its impact on grid-side waveform quality, a second-order RLC low-pass filter is implemented at the input side to provide an independent low-impedance flow path for the ripple current. This facilitates the analysis of the output current ripple locus in relation to various factors such as output/input switching sequences and converter regulation system influence [41].

The input current of the IMC is generated by chopping the load current, resulting in a waveform characterized by a sequence of intermittent current pulses. To ensure the stability of the input voltage, a parallel filter capacitor C_f is required at the input terminal of the IMC. Furthermore, a series filter inductor L_f is necessary to attenuate the transmission of high-frequency harmonic components of the input current between the power supply and the IMC input. As a result, the high-frequency harmonic current primarily flows through the low-impedance path provided by C_f, which forms a loop and reduces its impact on the waveform quality of the source. Collectively, C_f, L_f, along with a damping resistor R_f, constitute the input filter. A single-phase equivalent circuit of the IMC system is shown in Figure 5, where the filter is a typical two-port network.

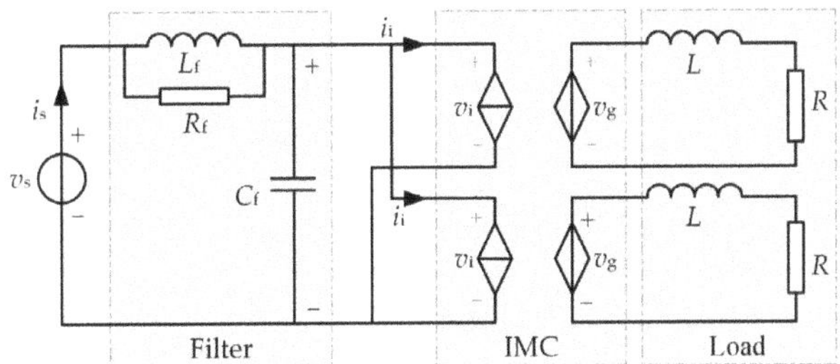

Figure 5. A double-phase equivalent of the IMC system.

Using transfer functions, the following analysis is performed on the two-port network in the complex frequency domain of the *s*-plane: The source voltage vs. and the MC input-side current i_i are selected as the independent variables of the port network; then, the source current i_s and the input-side voltage v_i can be expressed using the hybrid parametric equation (*h*-equation) as follows:

$$\begin{bmatrix} v_i(s) \\ i_s(s) \end{bmatrix} = \begin{bmatrix} h_{11} & h_{12} \\ h_{21} & h_{22} \end{bmatrix} \begin{bmatrix} -i_i(s) \\ v_s(s) \end{bmatrix} = \boldsymbol{h} \begin{bmatrix} -i_i(s) \\ v_s(s) \end{bmatrix} \tag{14}$$

The values of the elements of the hybrid parameter matrix $\boldsymbol{h}$ are as follows:

$$\begin{cases} h_{11} = \left.\dfrac{v_i(s)}{-i_i(s)}\right|_{v_s(s)=0} = \dfrac{2 \cdot sR_fL_f}{s^2R_fL_fC_f + sL_f + R_f} \\ h_{12} = \left.\dfrac{v_i(s)}{v_s(s)}\right|_{i_i(s)=0} = \dfrac{sL_f + R_f}{s^2R_fL_fC_f + sL_f + R_f} \\ h_{21} = \left.\dfrac{i_s(s)}{-i_i(s)}\right|_{v_s(s)=0} = -h_{12} \\ h_{22} = \left.\dfrac{i_s(s)}{v_s(s)}\right|_{i_i(s)=0} = \dfrac{(sL_f + R_f)sC_f}{s^2R_fL_fC_f + sL_f + R_f} \end{cases} \tag{15}$$

where L_f, C_f, and R_f are the filter inductance, capacitance, and damping resistance, respectively.

Appropriate filter parameters are a prerequisite for the reliable operation of the IMC.

The input voltages i_{iA}, i_{iB}, and i_{iC} result from the high-frequency modulation of the output current from the switching matrix. Furthermore, in the multiphase indirect matrix converter discussed in this paper, a low-pass filter is integrated between the input side and the grid side. This setup aims to curb the ingress of high-frequency harmonic components from the input current into the power grid. By doing so, it alleviates interference with other equipment at the common connection point and effectively filters out undesirable high-frequency harmonics.

The design of the filter necessitates careful consideration of its limitations and device selection criteria [30], which are evaluated based on the IMC equivalent circuit model illustrated in Figure 5. Initially, it is crucial to address the constraints regarding switching ripple and low-frequency harmonics. Since the fifth and seventh harmonics exhibit the highest amplitudes in the grid voltage, for simplification purposes, we will exclude the seventh harmonic from actual calculations. As depicted in Figure 5, the filter amplifies the harmonic component of the low-frequency voltage on the grid side. Therefore, it is imperative to restrict the low-frequency voltage gain coefficient of the filter to suppress low-frequency harmonics in the input voltage.

$$|G(j7\omega_b)| \leq 2\ \text{dB} \tag{16}$$

where w_b represents the input fundamental angular frequency. Similarly, to effectively suppress the ripple component of the current carrier frequency on the grid side, it is necessary to adjust the current gain coefficient of the filter at the carrier angular frequency w_c.

$$|G(j\omega_c)| \leq -26\ \text{dB} \tag{17}$$

Secondly, it is important to consider the limitations of inductor voltage drop and capacitor current. The ability to regulate the power factor is dependent on restricting the fundamental current of the filter capacitor. Therefore, it is crucial to take into account these limitations when designing the system for optimal performance.

$$I \approx \omega_b\, C_f\, v_s \leq 0.2 I_i^{\text{rated}} \tag{18}$$

where I_i^{rated} is the effective value of input current fundamental wave under rated load. To meet the voltage utilization requirements, it is necessary to limit the maximum fundamental voltage drop across the filtering inductance, under rated load conditions.

$$v \approx \omega_b L_f \sqrt{I^2 + \left(I_i^{\text{rated}}\right)^2} \leq 0.03 V_s \tag{19}$$

where V_s is the RMS value of the gride-side voltage. At the same time,

$$I_i^{\text{rated}} = \frac{P_i^{\text{rated}}}{3V_s} = \frac{\left(V_o^{\text{rated}}\right)^2 R_o}{3V_s} = \frac{(85\ \text{V})^2 \times 10\ \Omega}{3 \times 100\ \text{V}} \approx 8.76\ \text{A} \tag{20}$$

where P_i^{rated} is the rated active power, and V_o^{rated} is the RMS of the rated output voltage. From the above equation, the upper limit of the filter capacitance and inductance is as follows:

$$\begin{cases} C_f \leq 55.8\ \mu\text{F} \\ L_f \leq 1.09\ \text{mH} \end{cases} \tag{21}$$

Finally, the damping resistance R_f is carefully chosen to prevent oscillation in the CL filter. It is crucial to limit the maximum power loss of the damping resistor R_f.

$$3R_f {i_R}^2 \leq 0.01 P_i^{\text{rated}} \tag{22}$$

where i_R is the filter resistor current. Based on the above analysis, the main performance indicators considered in the filter design process are determined according to the required parameters and comprehensive practical considerations as follows:

1. The attenuation ratio for high-order harmonics should be no less than 26 dB;
2. The gain of the seventh and lower-order harmonics must be limited to 2 dB;
3. The voltage drop in the inductor must not exceed 3% of the fundamental amplitude;
4. Reactive current flowing into the capacitor shall not exceed 20% of the rated current;
5. The loss on the R_f is less than 1% of the rated power.

Similarly, the output voltage v_g and input current i_i are chosen as the independent variables of the port network, and the output current i_g and input voltage v_i can be expressed using a mixed parameter equation (h-equation).

$$\begin{bmatrix} i_g(s) \\ v_i(s) \end{bmatrix} = \begin{bmatrix} h_{11} & h_{12} \\ h_{21} & h_{22} \end{bmatrix} \begin{bmatrix} i_i(s) \\ v_g(s) \end{bmatrix} = \boldsymbol{h} \begin{bmatrix} i_i(s) \\ v_g(s) \end{bmatrix} \quad (23)$$

The values of the elements of the hybrid parameter matrix $\boldsymbol{h}$ are as follows:

$$\begin{cases} h_{11} = \left.\dfrac{v_i(s)}{-i_i(s)}\right|_{v_s(s)=0} = \dfrac{2 \cdot sR_fL_f}{s^3R_fL_f^2C_f + s^2(R_f^2L_fC_f + L_f^2) + 2sL_fR_f + R_f^2} \\ h_{12} = \left.\dfrac{v_i(s)}{v_s(s)}\right|_{i_i(s)=0} = \dfrac{1}{sL_f + R_f} \\ h_{21} = \left.\dfrac{i_s(s)}{-i_i(s)}\right|_{v_s(s)=0} = \dfrac{2sR_fL_f}{s^2R_fL_fC_f + sL_f + R_f} \\ h_{22} = \left.\dfrac{i_s(s)}{v_s(s)}\right|_{i_i(s)=0} = \dfrac{1}{q} \end{cases} \quad (24)$$

3.3. Current Ripple Model

As mentioned above, the SVM used in the IMC uses a limited number of active and zero vectors to generate a continuously changing target vector. The difference between the actual and targeted vectors results in the presence of numerous high-frequency harmonics in the input and output of the IMC. These harmonic components, which are filtered by the inductor/capacitor filter, manifest as output current ripples.

With the help of the Clarke transformation, the six-phase current quantities can be transformed into vector forms on the abc plane. The output current vector i_g can be represented as follows:

$$i_g = \frac{2}{3}\begin{bmatrix} 1 & 1e^{j2\pi/3} & e^{j4\pi/3} & e^{j\pi/6} & e^{j5\pi/6} & e^{j3\pi/2} \end{bmatrix} \cdot \begin{bmatrix} i_A & i_B & i_C & i_U & i_V & i_W \end{bmatrix}^T \quad (25)$$

where i_A, i_B, i_C, i_U, i_V, and i_W are the six-phase current on the output side, and the superscript T denotes matrix transpose. Similarly, the output voltage vector v_g, input current vector i_i, and input current vector v_i can be obtained. From this, it is easy to see that by generalizing the variables in (18) to vectors, the equation form remains unchanged.

Using small-signal analysis methods, the vector-expanded form of the nonlinear Equation (18) can be linearized. Thereafter, by defining small variations $\tilde{i}_g(s) = i_g(s) - \boldsymbol{I}_g(s)$, $\tilde{v}_g(s) = v_g(s) - \boldsymbol{V}_g(s)$, $\tilde{i}_i(s) = i_i(s) - \boldsymbol{I}_i(s)$, and $\tilde{v}_i(s) = v_i(s) - \boldsymbol{V}_i(s)$ around equilibrium points $\boldsymbol{I}_g(s)$, $\boldsymbol{V}_g(s)$, $\boldsymbol{I}_i(s)$, and $\boldsymbol{V}_i(s)$, and assuming that the ripple of input current $\tilde{i}_i = 0$ when the power supply is stable, it is possible to derive the small-signal model of the output side as follows:

$$\tilde{i}_g(s) = \frac{1}{sL_f + R_f}\tilde{v}_g(s) \quad (26)$$

Taking the inverse Laplace transform of (19), we can obtain the time-domain integral form of the output current ripple vector expression as follows:

$$\tilde{i}_g(s) = \frac{1}{L_f}\int_0^t \tilde{v}_g \, d\tau \tag{27}$$

Combined with the PWM switching sequence shown in Figure 3, the integration Equation (27) can be discretized as follows:

$$\tilde{i}_g = \frac{1}{L_f}\int_0^t \tilde{v}_g \, d\tau = \tilde{i}_{g,k} + \frac{\tilde{v}_{g,k}}{L_f}(t - t_{k-1}) \tag{28}$$

$$\tilde{i}_{g,k} = \tilde{i}_{g,k} + \frac{\tilde{v}_{g,k}}{L_f}T_k = \sum_{n=1}^{k} \frac{\tilde{v}_{g,\,k}}{L_f}T_n \tag{29}$$

where $k = 1, 2, \ldots, 14$, $t \in (t_{k-1}, t_k)$ (indicated as O, A, B...in Figure 3), $\tilde{i}_{g,k-1}$ is the output current ripple vector at $t = t_{k-1}$ and $\tilde{i}_{g,0} = 0$; $\tilde{v}_{g,k}$ is the output voltage error vector within (t_{k-1}, t_k); $T_k = t_k - t_{k-1}$.

The evaluation of the output current quality is typically conducted through quantification of the total harmonic distortion (THD). Nevertheless, the THD represents a frequency-domain metric that entails significant computational effort. The ripple constitutes a time-domain variable that incorporates all harmonic components of the output current. Consequently, the RMS value of the ripple signal can serve as an effective equivalent for characterizing the THD; that is, within a carrier cycle, the THD of the output current can be defined as follows:

$$THD|_{T_c} = \frac{\sqrt{\frac{1}{\pi}\int_0^{\pi} \tilde{i}^2_{g,rms}dt}}{i_g^{fund}} \times 100\% \tag{30}$$

where i_g^{fund} refers to the fundamental magnitude of the output current, and $\tilde{i}_g^{rms}$ represents the RMS of output current ripple within unit carrier frequency cycle T_c, which is defined as follows:

$$\tilde{i}_{g,rms} = \sqrt{\left(\tilde{i}_{gA}^{rms}\right)^2 + \left(\tilde{i}_{gB}^{rms}\right)^2 + \left(\tilde{i}_{gC}^{rms}\right)^2 + \left(\tilde{i}_{gU}^{rms}\right)^2 + \left(\tilde{i}_{gV}^{rms}\right)^2 + \left(\tilde{i}_{gW}^{rms}\right)^2} \tag{31}$$

where $\tilde{i}_{gA}^{rms}$, $\tilde{i}_{gB}^{rms}$, $\tilde{i}_{gC}^{rms}$, $\tilde{i}_{gU}^{rms}$, $\tilde{i}_{gV}^{rms}$, and $\tilde{i}_{gW}^{rms}$ denote the RMS of the six-phase source ripple currents. The maximum value of the ripple is limited to

$$\left|\tilde{i}_{g,max}\right| = \max\{|OA|, |OB| \ldots |OF|\} = \max(\left|\tilde{i}_{g,k}\right|) \tag{32}$$

where $\max(\left|\tilde{i}_{g,k}\right|)$ is the maximum value representing the output ripple vector, which is used to analyze the dynamic changes in the output current ripple locus.

3.4. Spread-Spectrum Modulation Based on Current Ripple Vectors

As depicted in Figure 4b, the locus of the output current ripple in the multiphase IMC is correlated with the amplitude and phase of the input–output voltages, among other factors. The fluctuation of the DC-link voltage v_{pn} directly determines the output ripple vector. The carrier frequency based on the dynamic variation in the ripple locus no longer fluctuates within a specific range and has unique advantages in the spread spectrum. Matching high carrier frequencies in the high ripple range can expand the energy concentrated at specific frequencies to a wider spectrum, effectively reducing the THD. Simultaneously, matching low carrier frequencies in the low ripple range improves system efficiency while achieving the spread spectrum.

By combining the two principles mentioned above, the correspondence between the ripple locus and carrier frequency is further established. From (22), it can be seen that the length of each segment of the locus is only related to the carrier period T_c. It can be inferred that the scale of the ripple can be easily manipulated by modulating the carrier frequency. Therefore, in this section, by dynamically changing the carrier frequency with the ripple locus, the variation is no longer based on a fixed magnitude. This method aims to optimize the output spectrum and enhance system operational efficiency.

The variable T_c appears in each term of the ripple expression in (22). Hence, (25) can be rewritten as follows:

$$\left|\widetilde{i}_{\mathrm{g,max}}\right| = \max(\left|\sum_{n=1}^{k} \frac{\widetilde{v}_{\mathrm{g,k}}}{L_{\mathrm{f}}}\right|)T_{\mathrm{n}} = \max(\left|\sum_{n=1}^{k} \frac{\widetilde{v}_{\mathrm{g,k}} d_{\mathrm{n}}}{L_{\mathrm{f}}}\right|)T_c \tag{33}$$

where k = 1, 2, ..., 14; $d_n = T_{\mathrm{n}}/T_{\mathrm{c}}$. This demonstrates that the output current ripple is proportional to the carrier period T_c.

Within each carrier cycle, the carrier frequency is modified to maintain the modulation index $\left|\widetilde{i}_{\mathrm{g,max}}\right|$ at its specified value $\left|\widetilde{i}_{\mathrm{g,max}}\right|^*$. It is evident that the carrier frequency $f_{\mathrm{c,carrier}}$ may be computed uniquely once the amplitude limit of the input voltage ripple vector in each cycle is known.

$$f_{\mathrm{c,carrier}} = \max(\left|\sum_{n=1}^{k} \frac{\widetilde{v}_{\mathrm{g,k}} d_n}{L_{\mathrm{f}}}\right|)/\left|\widetilde{i}_{\mathrm{i,\,max}}\right|^* \times 20\ \mathrm{kHz} \tag{34}$$

The RMS of the current ripple can be used in place of the THD as a standard indicator for assessing the quality of the current on the output side. Given that THD is a physical quantity that has been established over a long period of time, a suitable method must be used to assess if it complies with design criteria. In particular, it is sufficient to verify that the RMS of the output current ripple of any carrier period $\widetilde{i}_{\mathrm{g,rms}}$ satisfies the pertinent THD requirement after averaging over a certain time interval (typically equals to a fundamental period T_{b}).

$$THD = \frac{\frac{1}{T_b}\sqrt{\int_0^{T_b} \widetilde{i}^2_{\mathrm{g,rms}} dt}}{i_{\mathrm{g}}^{\mathrm{rate}}} \tag{35}$$

The traditional periodic PWM carrier frequencies fluctuate within a specific range. However, the carrier frequency defined by (19) no longer specifies a range of variation; instead, it dynamically adjusts with the trajectory of the ripple. This modification leads to a more standardized usage and eliminates the debugging process inherent in classical periodic PWM. It also facilitates clearer numerical indexes, thereby enhancing efficiency in carrier frequency fluctuation. Consequently, this approach can further reduce harmonic spikes near specific switching frequencies and promote a more uniform energy distribution within each frequency band.

4. Experimental Results

To assess the impact and effectiveness of the proposed CR-SSM method in terms of spreading spectrum, enhancing power quality, and improving efficiency for multiphase IMC systems, a 3 × 6-phase IMC system prototype was utilized for simulation and experimental validation. Furthermore, to further confirm the efficacy of the proposed CR-SSM method, it was compared with various conventional periodic PWM strategies such as triangular (TWPWM), sinusoidal (SWPWM), and exponential (EWPWM) periodic PWM.

The experimental prototype of the IMC platform is shown in Figure 6. The controller was a DSP (TI F28379D) + FPGA (Intel 5CEFA4) architecture, where the SVM modulation algorithm was completed independently by FPGA. And DSP was used for the online calculation of the carrier frequency. The rectifier circuit consisted of six Infineon FF200R12KT3 power modules, while the inverter circuit included two two-level IPM modules PM25CLA120 from the MITSUBISHI. Measurement equipment such as oscilloscopes and power analyzers (WT5000) were all from the Yokogawa corporation, Tokyo, Japan. The specific system parameters are shown in Table 2.

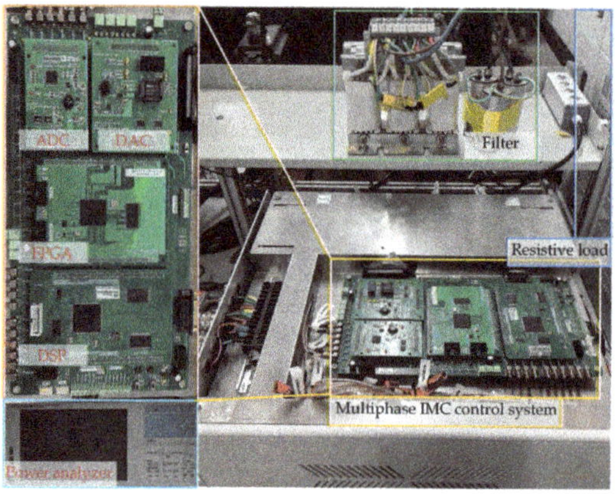

Figure 6. The experimental 3 × 6 IMC platform.

Table 2. Parameter settings.

Parameters	Symbol	Values
Effective value of grid-side phase voltage	V_s	100 V
Grid-side voltage frequency	f_s	50 Hz
Grid-side filter capacitors	C_f	35 μF
Grid-side filter inductors	L_f	1.37 mH
Grid-side filter resistors	R_f	10 Ω
Load resistance	R	10 Ω
Load inductance	L	30 mH
Output voltage frequency	f_o	20 Hz
Fixed carrier frequency	f_{c0}	10 kHz

4.1. Analysis of Harmonic Suppression

To demonstrate the change in carrier frequency in terms of distribution characteristics, with respect to the input voltage phase angle $\widetilde{\theta}_i$, the scalar value of f_c was assigned the same phase information as $\widetilde{\theta}_i$ and converted to the vector form, i.e.,

$$f_c = f_c\, e^{j\theta_i} \tag{36}$$

The waveform of the vectorized carrier frequency f_c can be represented using a polar coordinate system. In this system, the radial distance represents the value of the carrier frequency f_c, and the angular coordinate represents the phase angle $\widetilde{\theta}_i$ of the input voltage vector. The orange and purple circles in Figure 7a, respectively, show the carrier frequency waveform for the CR-SSM method when $f_c = f_{c,max}$ = 20 kHz and $f_c = f_{c,min}$ = 5 kHz. The annular region between the two circles represents the feasible range of carrier frequency modulation. In order to visualize the distribution of the modulated carrier frequency in each frequency band, Figure 7b illustrates the histogram of the output harmonic distribution under the CR-SSM method in the 5–20 kHz band. The abscissa represents the harmonic frequency band, and it is divided into 17 equal segments to show the actual output of 5–20 k. The ordinate represents the approximate harmonic amplitude distribution in this band.

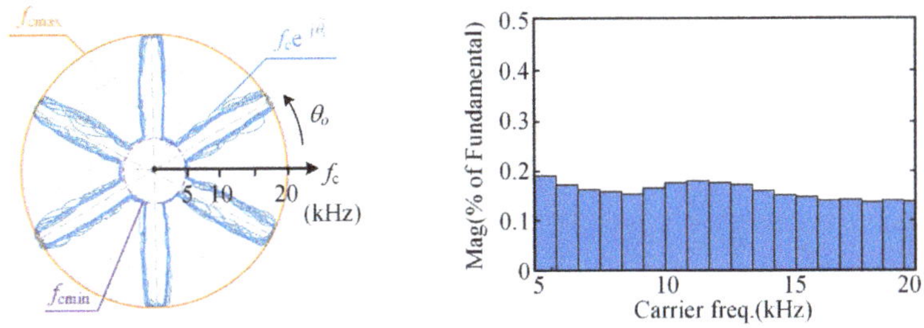

(a) (b)

Figure 7. Plot of carrier frequency with CR-SSM of (**a**) vectorized waveform and (**b**) distribution histogram.

A comparison of three types of periodic PWM with the proposed CR-SSM method was performed for further evaluation of the effectiveness of the proposed method. The vectorized waveform of the frequency-modulated wave and the harmonic spectrum near 10 k for the other three types of periodic PWM methods are shown in Figure 8. The following observations can be made:

1. The strategy of periodic PWM to reduce the carrier frequency always results in a lower range of carrier frequencies compared to the CR-SSM method while maintaining the same limit on output current ripple. It can be seen that the classical periodic PWM still has harmonic spikes near the carrier frequency of 10 kHz, the variable band energy is still more concentrated, and the THD value is higher. The proposed CR-SSM method has better spectral characteristics, no harmonic spikes at specific frequencies, uniform high-frequency energy distribution, and better THD values.
2. The carrier frequency waveform displays periodicity along the direction of the output voltage vector angle $\widetilde{\theta}_o$, but its locus is affected by the input frequency $\widetilde{\theta}_i$, and it does fully coincide with the previous cycle after a complete cycle ($\widetilde{\theta}_o = 2\pi$).

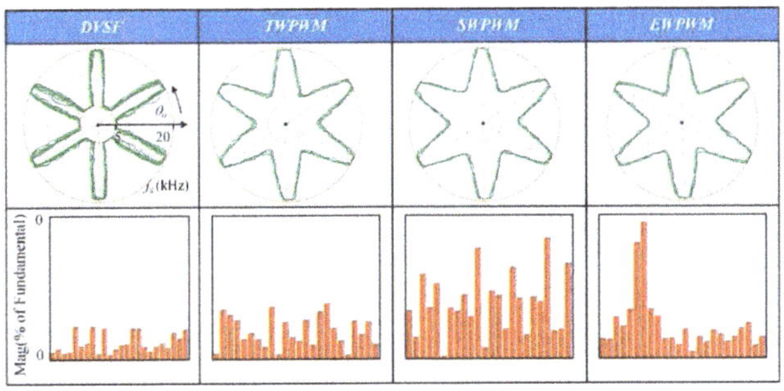

Figure 8. Comparison of the frequency variation locus between classic periodic PWM and CR-SSM switching method.

4.2. Experimental Analysis of the Converter Efficiency

As the carrier frequency of the IMC system decreases, so do the losses. The efficiency of IMC can be improved by modulating the carrier frequency with CR-SSM, which employs a wider range of carrier frequency variations compared to the classical periodic PWM method. Therefore, it is necessary to conduct further experimental verification to determine the effect of CR-SSM and periodic PWM methods on the efficiency of IMC systems. Figure 9 shows the operational efficiency of the system under different methods. A comprehensive analysis of the data reveals the following observations:

1. The efficiency of the CR-SSM method is always lower than that of the periodic PWM method at different voltage transmission ratios q;
2. As the voltage transfer ratio q increases, the efficiency of the system also increases.

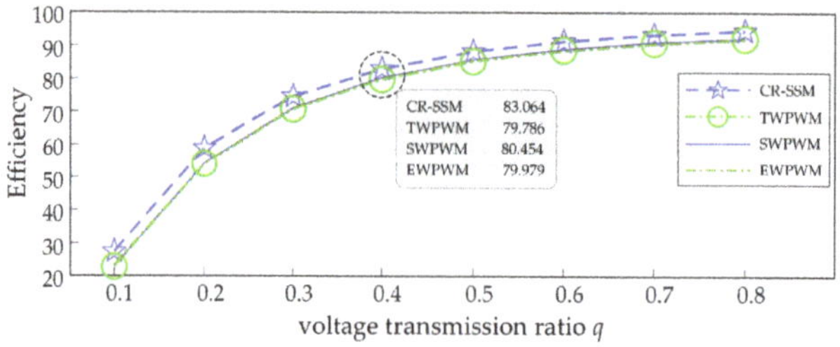

Figure 9. Operating efficiency of various methods under different working conditions.

Figure 9 illustrates the three periodic PWMs (EWPWM, TWPWM, and SWPWM) and the operational efficiency of the multiphase IMC system under the proposed CS-SSM strategy. As depicted, the system efficiency of the CS-SSM method escalates from 27.5% to 94.5%; in SWPWM, it increases from 23.1% to 92.1%; in EWPWM, it increases from 23.3% to 91.9%; and in TWPWM, it increases from 22.8% to 92.2%. Notably, at modulation 0.8, the CR-SSM method boosts efficiency by 2.4%, 2.3%, and 2.6% compared to EWPWM, TWPWM, and SWPWM, respectively. Similarly, at modulation 0.6, the CR-SSM method enhances efficiency by 2.7%, 2.3%, and 2.9% compared to EWPWM, TWPWM, and SWPWM, respectively. The conjunction of these findings with Figure 9 demonstrates that the proposed CR-SSM method not only ensures a spread-spectrum effect but also enhances the operational efficiency of the multiphase IMC system by leveraging the carrier frequency density function based on the current ripple.

4.3. Experimental Results on Input–Output Waveform Quality

Figure 10 illustrates a comparison of the total harmonic distortion (THD) of load current under the CR-SSM method and periodic PWM. It can be observed that, under different voltage transfer ratio q conditions, the THD of the load current under the proposed CR-SSM method is lower than that of the periodic PWM method. This indicates that the carrier frequency modulation technique can effectively reduce the harmonic level of the IMC load current. Furthermore, as the modulation depth increases, the THD of the proposed CR-SSM is always lower than that of the periodic PWM method.

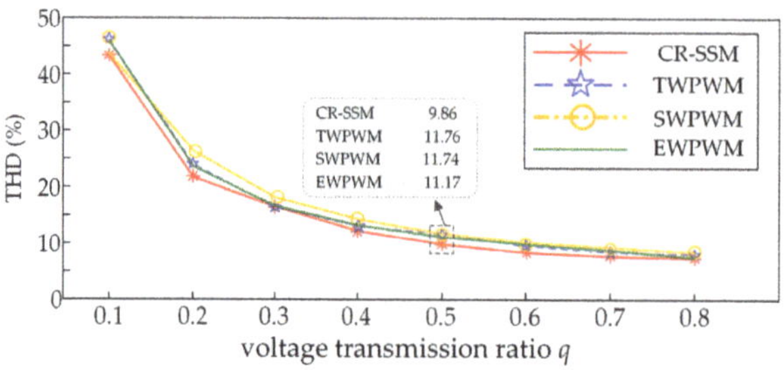

Figure 10. THD comparison of CR-SSM and three classical periodic PWM methods.

As depicted in Figure 10, the total harmonic distortion (THD) of the output current diminishes with increasing modulation. Notably, employing the proposed CR-SSM strategy results in a significant reduction in THD from 41.34% to 7.32%. Similarly, the THD for the TWPWM method decreases from 41.34% to 7.65%, while for the SWPWM method, it decreases from 43.09% to 7.89%, and for the EWPWM method, it decreases from 43.75% to

7.67%. At a modulation index of 0.5, the THD for CR-SSM is 9.86%, while for TWPWM, SWPWM, and EWPWM, it stands at 11.76%, 11.74%, and 11.17%, respectively. This trend underscores that, across all modulation methods, the THD of the proposed CR-SSM strategy remains consistently lower than the other three periodic PWM methods. The findings from Figures 4–8 corroborate that the proposed CR-SSM strategy enhances the output current of the multiphase IMC system while ensuring the desired spread-spectrum effect.

The time-domain waveforms of the source current, DC-link voltage, input line voltage, and load current under various methods are shown in Figure 11. Figure 12 depicts the dynamic experimental waveforms using four different methods as the output frequency varies from 20 Hz to 40 Hz. It can be observed from Figures 11 and 12 that the proposed CR-SSM method ensures a balanced sinusoidal three-phase source current, DC-link voltage, and load current without significant fluctuations. Moreover, it effectively suppresses harmonic amplitude at switching frequencies.

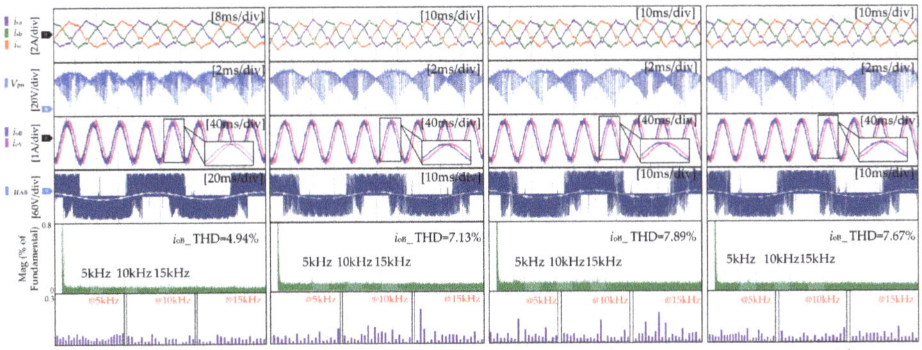

(a) (b) (c) (d)

Figure 11. Comparison between the proposed CR-SSM and classic periodic PWM: (**a**) CR-SSM; (**b**) TWPWM; (**c**) SWPWM; (**d**) EWPWM.

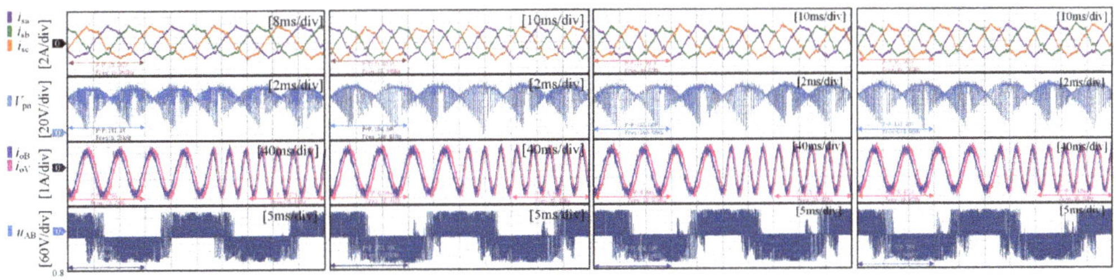

(a) (b) (c) (d)

Figure 12. Comparison between the proposed CR-SSM and classic periodic PWM dynamics: (**a**) CR-SSM; (**b**) TWPWM; (**c**) SWPWM; (**d**) EWPWM.

5. Conclusions

This study analyzes output current ripple to construct a carrier frequency density function and proposes a CR-SSM method suitable for multiphase IMC systems. The method selectively modulates carrier frequency without requiring changes to other modules, thus serving as a complementary tool to optimize PWM modulators like SVM and carrier modulation. CR-SSM utilizes output current ripple as an indicator to build a switch frequency density function, addressing the limitations of traditional periodic PWM in center frequency control. It offers a standardized application, eliminates the debugging process associated with classic periodic PWM, and defines clear numerical indicators. Additionally, it improves switch frequency efficiency, reducing harmoni2c peaks near specific switch frequencies. This promotes a more even energy distribution across frequency bands and optimizes system efficiency by correlating switch frequencies with output ripple heights. By distributing the power spectrum over a broader frequency range, the system's output current spectrum is optimized. Experimental results demonstrate that compared to periodic PWM, the CR-SSM strategy effectively reduces system total harmonic distortion (THD) while enhancing operational efficiency in achieving spread-spectrum operation modes.

Author Contributions: Conceptualization, Z.Z.; data curation, L.X.; formal analysis, L.X.; funding acquisition, Z.Z.; investigation, C.L.; methodology, L.X.; project administration, Z.Z.; resources, Z.Z.; software, Z.Z.; supervision, Z.Z. and Q.G.; validation, L.X.; visualization, L.X.; writing—original draft preparation, L.X. All authors have read and agreed to the published version of the manuscript.

Funding: This research was funded by the "Zhejiang Provincial Natural Science Foundation of China under Grant No. LY24E070001".

Data Availability Statement: The original contributions presented in the study are included in the article, further inquiries can be directed to the corresponding author.

Conflicts of Interest: The authors declare no conflicts of interest.

References

1. Wang, R.; Lin, Z.; Du, J.; Wu, J.; He, X. Direct sequence spread spectrum-based PWM strategy for harmonic reduction and communication. *IEEE Trans. Power Electron.* **2016**, *32*, 4455–4465. [CrossRef]
2. Raghuram, M.; Chauhan, A.K.; Singh, S.K. High-Gain-Integrated Switched Capacitor Indirect Matrix Converter. *IEEE J. Emerg. Sel. Top. Power* **2019**, *3*, 1846–1853. [CrossRef]
3. Shi, T.; Wang, Z.; Ma, C.; Yan, Y.; Song, P. A Variable Switching Frequency PWM Method for Indirect Matrix Converters. In Proceedings of the 10th International Conference on Power Electronics, Machines and Drives (PEMD 2020), Online Conference, 15–17 December 2020; pp. 943–948.
4. Padhee, V.; Sahoo, A.K.; Mohan, N. Modulation techniques for enhanced reduction in common-mode voltage and output voltage distortion in indirect matrix converters. *IEEE Trans. Power Electron.* **2016**, *32*, 8655–8670. [CrossRef]
5. Nguyen, H.N.; Lee, H.H. An effective SVM method for matrix converters with a superior output perfor-mance. *IEEE Trans. Ind. Electron.* **2017**, *65*, 6948–6958. [CrossRef]
6. Empringham, L.; Kolar, J.W.; Rodriguez, J.; Wheeler, P.W.; Clare, J.C. Technological Issues and Industrial Application of Matrix Converters: A Review. *IEEE Trans. Ind. Electron.* **2013**, *60*, 4260–4271. [CrossRef]
7. Chen, J.; Sha, D.; Zhang, J. Current ripple prediction and DPWM-based variable carrier frequency control for full ZVS range three-phase inverter. *IEEE Trans. Ind. Electron.* **2020**, *68*, 1412–1422. [CrossRef]
8. Casadei, D.; Serra, G.; Tani, A.; Zarri, L. Optimal use of zero vectors for minimizing the output current distortion in matrix converters. *IEEE Trans. Ind. Electron.* **2008**, *56*, 326–336. [CrossRef]
9. Casadei, D.; Serra, G.; Tani, A.; Zarri, L. Theoretical and experimental analysis for the RMS current ripple minimization in induction motor drives controlled by SVM techniqu. *IEEE Trans. Ind. Electron.* **2004**, *51*, 1056–1065. [CrossRef]
10. Attia, H.; Freddy, T.; Che, H.; Hew, W.; Elkhateb, A. Confined band variable carrier frequency pulse width modulation (CB-VSF PWM) for a single-phase inverter with an LCL filter. *IEEE Trans. Power Electron.* **2016**, *32*, 8593–8605. [CrossRef]
11. Qin, X.; Zhou, B.; Wei, J.; Lei, J.; Liu, X.; Han, N. Distortion Analysis and Duty Ratio Correction Algorithm for Asymmetric Modulation of Two-Stage Matrix Converter. *IEEE Trans. Ind. Electron.* **2014**, *62*, 351–362. [CrossRef]
12. Xie, L.; Ruan, X.; Zhu, H.; Lo, Y.-K. Common-Mode Voltage Cancellation for Reducing the Common-Mode Noise in DC–DC Converters. *IEEE Trans. Ind. Electron.* **2021**, *5*, 3887–3897. [CrossRef]
13. Onederra, O.; Kortabarria, I.; De Alegria, I.M.; Andreu, J.; Gárate, J.I. Three-Phase VSI Optimal Switching Loss Reduction Using Variable Carrier frequency. *IEEE Trans. Power Electron.* **2016**, *32*, 6570–6576. [CrossRef]
14. Marouani, K.; Baghli, L.; Hadiouche, D.; Kheloui, A.; Rezzoug, A. A New PWM Strategy Based on a 24-Sector Vector Space Decomposition for a Six-Phase VSI-Fed Dual Stator Induction Motor. *IEEE Trans. Ind. Electron.* **2008**, *5*, 1910–1920. [CrossRef]
15. Gamoudi, R.; Elhak Chariag, D.; Sbita, L. A Review of Spread-Spectrum-Based PWM Techniques—A Novel Fast Digital Implementation. *IEEE Trans. Power Electron.* **2018**, *12*, 10292–10307. [CrossRef]
16. Cano, J.M.; Navarro-Rodriguez, A.; Suarez, A.; Garcia, P. Variable Carrier frequency Control of Distributed Resources for Improved System Efficiency. *IEEE Trans.* **2018**, *54*, 4612–4620.
17. Jamal, H.; Albatran, S.; Smadi, I. Variable carrier frequency algorithm for optimal tradeoff between switching losses and total demand distortion in grid-tied three-phase voltage-source inverter. In Proceedings of the 2016 IEEE Energy Conversion Congress and Exposition (ECCE), Milwaukee, WI, USA, 18–22 September 2016; IEEE: Piscataway, NJ, USA, 2016; pp. 1–7.
18. Jiang, D.; Wang, F. Variable carrier frequency PWM for three-phase converters based on current ripple prediction. *IEEE Trans. Power Electron.* **2013**, *28*, 4951–4961. [CrossRef]

19. Atamaniuk, V.; Kuznyetsov, O.; Lunkova, H. Driving the PWM-Controlled System to Monostable Behavior by Periodic Modulation of a Carrier Signal. In Proceedings of the 2022 IEEE 4th International Conference on Modern Electrical and Energy System (MEES), Kremenchuk, Ukraine, 20–22 October 2022; pp. 1–6.
20. Shan, Y.; Pei, X.; Sun, T.; Zhang, M.; Zhou, P.; Jiang, D. Space spread-spectrum strategy for MMC to reduce the conducted EMI. *IEEE Trans. Ind. Electron.* **2021**, *69*, 10807–10818. [CrossRef]
21. Chen, J.; Jiang, D.; Shen, Z.; Sun, W.; Fang, Z. Uniform distribution pulsewidth modulation strategy for three-phase converters to reduce conducted EMI and switching loss. *IEEE Trans. Ind. Electron.* **2019**, *67*, 6215–6226. [CrossRef]
22. Li, Q.; Jiang, D.; Zhang, Y. Analysis and calculation of current ripple considering inductance saturation and its application to variable carrier frequency PWM. *IEEE Trans. Ind. Electron.* **2019**, *34*, 12262–12273.
23. Chen, J.; Jiang, D.; Sun, W.; Shen, Z.; Zhang, Y. A Family of Spread-Spectrum Modulation Schemes Based on Distribution Characteristics to Reduce Conducted EMI for Power Electronics Converters. *IEEE Trans. Ind. Applicat.* **2020**, *56*, 5142–5157. [CrossRef]
24. Le, D.; Choi, S. Network CM EMI Reduction Using Sinusoidal Frequency Modulated Carrier Wave Indexing. In Proceedings of the 2023 11th International Conference on Power Electronics and ECCE Asia (ICPE 2023-ECCE Asia), Jeju Island, Republic of Korea, 22–25 May 2023; Volume 38, pp. 3321–3326.
25. Lumbreras, D.; Zaragoza, J.; Berbel, N.; Mon, J.; Gálvez, E.; Collado, A. Comprehensive Analysis of Hexagonal Sigma-Delta Modulations for Three-Phase High-Frequency VSC Based on Wide-Bandgap Semiconductors. *IEEE Trans. Power Electron.* **2021**, *36*, 7212–7222. [CrossRef]
26. Menon, M.A.; Jacob, B. A New Space Vector Pulse Density Modulation Scheme for Two-Level Five Phase Induction Motor Drive. *IEEE Trans. Ind. Electron.* **2022**, *69*, 12077–12085.
27. Huang, Y.; Xu, Y.; Zhang, W.; Zou, J. Modified Single-Edge SVPWM Technique to Reduce the Switching Losses and Increase PWM Harmonics Frequency for Three-Phase VSIs. *IEEE Trans. Power Electron.* **2020**, *35*, 10643–10653. [CrossRef]
28. Ji, Z.; Cheng, S.; Li, X.; Lv, Y.; Wang, D. An Optimal Periodic Carrier Frequency PWM Scheme for Suppressing High-Frequency Vibrations of Permanent Magnet Synchronous Motors. *IEEE Trans. Power Electron.* **2023**, *38*, 13008–13018. [CrossRef]
29. Ji, Z.; Cheng, S.; Ren, Q.; Li, X.; Lv, Y.; Wang, D. The Effects and Mechanisms of Periodic-Carrier-Frequency PWM on Vibrations of Multiphase Permanent Magnet Synchronous Motors. *IEEE Trans. Power Electron.* **2023**, *38*, 8696–8706. [CrossRef]
30. Chen, N.; Yu, S.; Gao, L.; Huang, Y.; Chen, R. Suppressing Interference Peak in an Active Power Filter via Periodic Carrier Frequency Modulation Based on a Spectrum Analysis Approach. *IEEE Trans. Electromagn. Compat.* **2019**, *61*, 1760–1770. [CrossRef]
31. Xu, J.; Nie, Z.; Zhu, J. Characterization and Selection of Probability Statistical Parameters in Random Slope PWM Based on Uniform Distribution. *IEEE Trans. Power Electron.* **2021**, *1*, 1184–1192. [CrossRef]
32. Ma, Y.; Yang, M.; Lyu, Z.; Li, S.; Xu, D. Application of Periodic Carrier Frequency Modulation for Harmonic and EMI Reduction in GaN-Based Motor Drive. In Proceedings of the 2020 23rd International Conference on Electrical Machines and Systems (ICEMS), Hamamatsu, Japan, 24–27 November 2020; pp. 778–782.
33. Nguyen, T.D.; Lee, H.-H. Dual Three-Phase Indirect Matrix Converter With Carrier-Based PWM Method. *IEEE Trans. Power Electron.* **2014**, *2*, 569–581. [CrossRef]
34. Benson, M.; Dornala, A.; Andleeb, M.; Lee, K.; Lee, W. Simple and Robust Carrier-Based PWM Technique for Single-Stage Three-phase Rectifier Indirect Matrix Converter. In Proceedings of the 2024 IEEE Applied Power Electronics Conference and Exposition (APEC), Long Beach, CA, USA, 25–29 February 2024; pp. 691–697.
35. Park, H.P.; Kim, M.; Jung, J.H. Spread spectrum technique to reduce EMI emission for an LLC resonant converter using a hybrid modulation method. *IEEE Trans. Power Electron.* **2017**, *33*, 3717–3721. [CrossRef]
36. Dasgupta, A.; Sensarma, P. Filter design of direct matrix converter for synchronous applications. *IEEE Trans. Ind. Electron.* **2014**, *61*, 6483–6493. [CrossRef]
37. Qiu, Z.; Chen, Y.; Kang, Y.; Liu, X.; Gu, F. Investigation into periodic signal-based dithering modulations for suppression sideband vibro-acoustics in PMSM used by electric vehicles. *IEEE Trans. Energy Convers.* **2020**, *36*, 1787–1796. [CrossRef]
38. Leoncini, M.; Bertolini, A.; Melillo, P.; Gasparini, A.; Levantino, S.; Ghioni, M. Spread-spectrum frequency modulation in a DC/DC converter with time-based control. *IEEE Trans. Power Electron.* **2022**, *38*, 4207–4211. [CrossRef]
39. Kumar, A.B.; Narayanan, G. Variable-Carrier frequency PWM Technique for Induction Motor Drive to Spread Acoustic Noise Spectrum with Reduced Current Ripple. *IEEE Trans. Ind. Appl.* **2016**, *52*, 3927–3938. [CrossRef]
40. Li, Q.; Jiang, D. Variable Carrier frequency PWM Strategy of Two-Level Rectifier for DC-Link Voltage Ripple Control. *IEEE Trans. Power Electron.* **2017**, *33*, 7193–7202. [CrossRef]
41. Xu, C.; Lu, S. Practical Online Modulation Method for Current Ripple and Switching Losses Reduction in the Three-Phase Voltage Source Inverters. *IEEE Trans. Power Electron.* **2021**, *2*, 1475–1490. [CrossRef]

energies

MDPI

Article

A DCM-Based Non-Isolated Step-Down DC Transformer

Minseung Kim [1], Donghee Choi [2] and Soo Hyoung Lee [3,*]

1 Department of Electrical Engineering, Mokpo National University, Muan 58554, Republic of Korea
2 Division of Converged Electronic Engineering, Cheongju University, Cheongju 28503, Republic of Korea; heechoi@cju.ac.kr
3 Division of Electrical, Electronic, and Control Engineering, Kongju National University, Cheonan 31080, Republic of Korea
* Correspondence: lsh@kongju.ac.kr

Abstract: DC transformers have emerged as essential devices for medium voltage DC (MVDC)-low voltage DC (LVDC) distribution systems. However, conventional step-down single-level converters have limits on the voltage level of the MVDC-LVDC distribution system. This paper proposes a non-isolated step-down (NISD) DC transformer based on discontinuous conduction mode (DCM). The proposed structure can withstand high voltage levels by sharing voltages between energy storage modules dividing voltage levels. The proposed NISD DC transformer determines operational modes based on energy storage modules and performs the voltage conversion process. The effectiveness of the proposed NISD DC transformer is verified based on a case study using a power system computer-aided design and electromagnetic transient simulation engine including DC (PSCAD/EMTDC™).

Keywords: DC distribution; DC transformer; modular multilevel converter; non-isolated converter; step-down converter

1. Introduction

As climate change poses increasingly significant environmental challenges, there is a worldwide increase in the demand for renewable sources of energy that are sustainable and environmentally friendly. In accordance with the International Energy Agency, renewable energy will comprise 29.8% of the world's overall electricity production by 2022 [1]. The growth of direct current (DC) power generation is driving the increase in renewable energy production, along with the rise of DC loads such as electric vehicle charging stations and data centers [2–4]. This attention is leading to a focus on DC power transmission and distribution [2–7]. A critical challenge in this area is the design of converters to reduce the voltage level from medium-voltage DC (MVDC) to low-voltage DC (LVDC) without using the bulky and expensive AC transformers that are currently in use. A DC transformer, commonly known as a DC-DC converter, uses switching components and energy storage implements for voltage transformation [8,9]. However, conventional single-stage step-down converters require high-rated voltages for each component, which leads to higher costs, larger sizes, and higher losses. As a consequence, the MVDC to LVDC distribution voltage levels are limited [10–12].

To overcome the disadvantages of replacing AC transformers, alternative isolated DC converters using medium- or high-frequency transformers have been proposed instead of the traditional bulky low-frequency transformers [13–16]. Despite this solution, these converters still exhibit certain drawbacks, such as losses during the DC-AC-DC conversion and the continued presence of AC transformers [17].

This paper proposes a non-isolated step-down (NISD) DC transformer based on discontinuous conduction mode (DCM). Currently, only DCM operation is considered, and the proposed structure divides high voltage using an energy storage module (ESM) comprised of series. The proposed NISD DC transformer determines operational mode based on the ESMs and performs voltage transformation, providing the following advantages:

(1) It is possible to achieve a high voltage ratio with a small number of components and a simple control method.
(2) It operates in DCM, reducing switching losses and improving efficiency.
(3) There is no requirement for an isolation transformer or galvanic isolation, which results in reduced cost and size.

The proposed NISD DC transformer uses capacitors to store electric energy in ESMs [18–27], unlike the conventional DC converter that uses inductors to store magnetic energy and perform voltage conversion [3,6,15,16,18,19,28–32]. Also, conventional flying capacitor multilevel converters have complex control schemes [24–26] in contrast to this structure, enabling flexible control of the output voltage by adjusting the charge and discharge duty of ESMs. The high voltage of MVDC is dispersed among ESMs, allowing for a reduction of each circuit element's rated voltage to the level of commonly used components. Currently, soft switching is considered for the only main switches excluding ESMs.

The remaining parts of this paper are structured as follows: Section 2 presents the topology and configuration of the proposed NISD DC transformer, as well as its operational modes and control strategy. Section 3 analyzes the IGBT voltage stress and current as well as the considerations for the inductor, ESM, and output capacitor during the design of the proposed NISD DC transformer. Section 4 provides the results of simulations conducted using a power system computer-aided design and electromagnetic transient including DC (PSCAD/EMTDC) to validate the effectiveness of the proposed NISD DC transformer in two cases: the steady state and with disturbances. Finally, Section 5 summarizes the key contributions of this paper and proposes several future research directions.

2. NISD DC Transformer Based on DCM

2.1. NISD DC Transformer Topology

The topology of the non-isolated step-down DC transformer is presented in Figure 1. The proposed topology operates based on capacitors in contrast to the conventional one that uses an inductor. It includes n energy storage modules for voltage transformation and operates according to a six-duty ratio segment mode. Each energy storage module is composed of a capacitor, marked as (C), and two insulated gate bipolar transistors (IGBTs), labeled as (S_a, S_b), which are charged and discharged by separate groups of IGBTs, marked as (S_{11}, $\cdots$, $S_{1\alpha}$ and S_{21}, $\cdots$, $S_{2\alpha}$). Moreover, the addition of the inductor (L) and the output capacitor (C_{out}) restrict excessive current and reduce the ripple in output voltage, respectively.

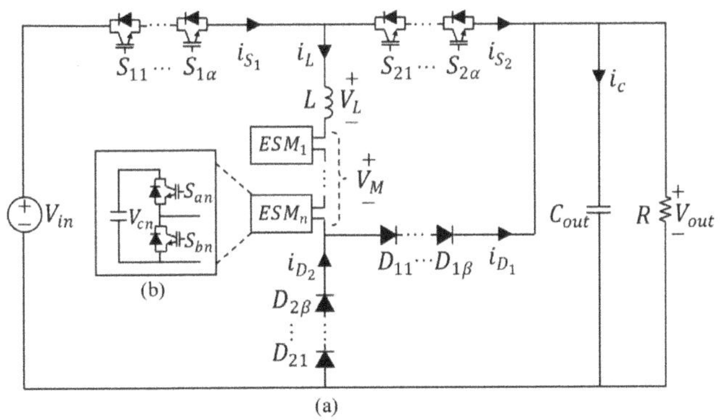

Figure 1. (**a**) NISD DC transformer topology including (**b**) ESM.

2.2. Operational Modes

The operational modes of the NISD DC transformer are composed of duty cycles divided into six segments ($M_1 - M_6$) based on the charging and discharging of the inductor and ESM. The operational sequence depicted in Figure 2 is demonstrated to operate in a steady state. The waveforms of voltage and current are periodic, with identical values at the beginning (t_0) and end (t_6) of each cycle. Figure 3 illustrates the NISD DC transformer's operational mode for each segment. And its current path is shown as red highlighted.

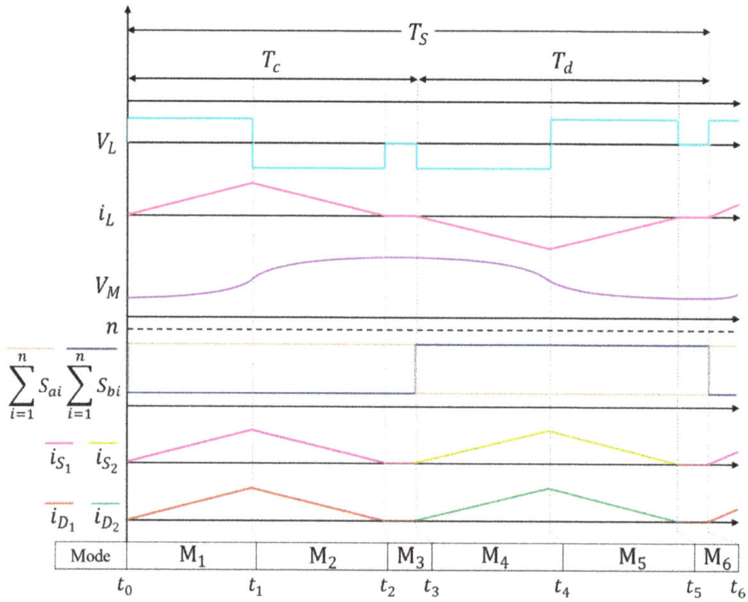

Figure 2. Operational sequences of the proposed DC transformer.

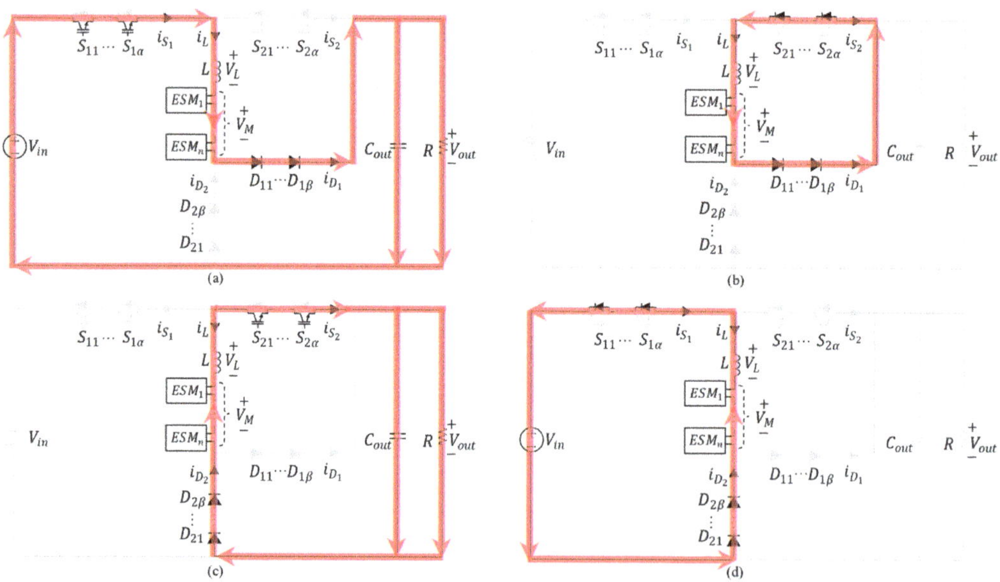

Figure 3. Operational modes of NISD DC transformer: **(a)** M_1, **(b)** M_2, **(c)** M_4, **(d)** M_5.

Before analyzing each operational mode, it is assumed that the IGBTs, diodes, and ESMs' capacitor balancing are ideal during operation. Moreover, the inductor has low inductance to make the inductor current periodically zero, and each capacitor is assumed to have a large enough capacitance to maintain a constant voltage at the output voltage V_{out} and the voltage V_{cn} ($n \in \mathbb{N}$) of each ESM.

The voltage reference of the ESM capacitor is defined by the relationship between the input voltage and the number of ESMs, as shown in (1).

$$V_c = \frac{V_{in}}{n} \tag{1}$$

The average of each ESM capacitor voltage is expressed as (2), and the operational mode is determined by the relationship between (1) and (2).

$$\overline{V_c} = \frac{1}{n}\sum_{i=1}^{n} V_{ci} \tag{2}$$

2.2.1. First Operational Mode $\mathrm{M_1} : (\mathrm{t_0} - \mathrm{t_1})$

During the steady state, the state at t_0 is identical to the state at t_6 in the prior operational sequence., i.e., both groups of IGBTs are off and the inductor current is zero. Afterward, IGBT group 1 ($S_{11}, \cdots, S_{1\alpha}$) is turned on, which causes the capacitor of the ESM to charge and the voltage to increase through the path, as shown in Figure 3a. During this process, the inductor prevents an overcurrent by filling the difference between the input voltage and the controlled ESM voltage and V_M, as shown in (3).

$$V_L = V_{in} - V_M - V_{out} \tag{3}$$

2.2.2. Second Operational Mode $\mathrm{M_2} : (\mathrm{t_1} - \mathrm{t_2})$

During the first operational mode, when the voltage charged to each ESM is sufficiently high to satisfy $\overline{V_c} > V_c$, the IGBT group 1 ($S_{11}, \cdots, S_{1\alpha}$) is turned off at t_1. At this moment, as illustrated in Figure 3b, current that has been charged to the inductor by circulating through the antiparallel diodes of IGBT group 2 ($S_{21}, \cdots, S_{2\alpha}$), the ESM, and diode group 1 ($D_{11}, \cdots, D_{1\beta}$) delivers energy to ESMs and then discharges to zero. At this time, the voltage of each ESM capacitor is increased by the ESM capacitor's voltage ripple ΔV_c to the reference voltage in (1) to $V_c + \Delta V_c$. The inductor voltage during discharge in this mode is determined by (4).

$$V_L = -V_M \tag{4}$$

2.2.3. Third Operational Mode $\mathrm{M_3} : (\mathrm{t_2} - \mathrm{t_3})$

As the load on the DC transformer approaches its maximum rated load, which is $\mathrm{M_3} = 0$ at full load, the duration of the third mode is reduced. However, its changes do not affect the stability of the output voltage considering the fixed switching frequency.

2.2.4. Fourth Operational Mode $\mathrm{M_4} : (\mathrm{t_3} - \mathrm{t_4})$

In the third operational mode, after the inductor current is completely discharged and IGBT group 2 ($S_{21}, \cdots, S_{2\alpha}$) turns on at t_3, the ESM capacitor discharges via the path shown in Figure 3c and the voltage decreases. As in the first mode, the inductor prevents overcurrent by covering the gap between the controlled ESM voltage and the output voltage, as given by (5).

$$V_L = V_{out} - V_M \tag{5}$$

2.2.5. Fifth Operational Mode $\mathrm{M_5} : (\mathrm{t_4} - \mathrm{t_5})$

After the ESM is discharged in the fourth mode and the voltage is sufficient to satisfy $\overline{V_c} < V_c$, IGBT group 2 ($S_{21}, \cdots, S_{2\alpha}$) is turned off at t_4. At this point, the inductor current that has been accumulated is diverted back to the input side and discharged to zero through the anti-parallel diodes of IGBT group 2 ($S_{21}, \cdots, S_{2\alpha}$), the ESM, and diode group 2 ($D_{21}, \cdots, D_{2\beta}$), as seen in Figure 3d. At this time, the ESM capacitors are continuously being discharged to reduce the voltage to $V_c - \Delta V_c$ according to the previous mode, and the voltage of the discharged inductor is equal to (6).

$$V_L = V_{in} - V_M \tag{6}$$

2.2.6. Sixth Operational Mode $M_6 : (t_5 - t_6)$

Similar to the third operational mode, when the DC transformer operates at full rated load, then $M_6 = 0$. The duration of the sixth operational mode is also longer when operating at lower load.

2.3. Control Strategy

The control logic of the NISD DC transformer is presented in Figure 4. Thereafter, it proceeds in sequence through the six operational modes detailed in Section 2.2. At the outset, the voltage of the ESM begins at zero; therefore, IGBT group 1 ($S_{11}, \cdots, S_{1\alpha}$) is turned on and operates as M_1. Subsequently, IGBT group 1 ($S_{11}, \cdots, S_{1\alpha}$) and IGBT group 2 ($S_{21}, \cdots, S_{2\alpha}$) are controlled by the voltage of ESMs, and the charging and discharging of ESMs are determined according to the six operational modes described in Section 2.2.

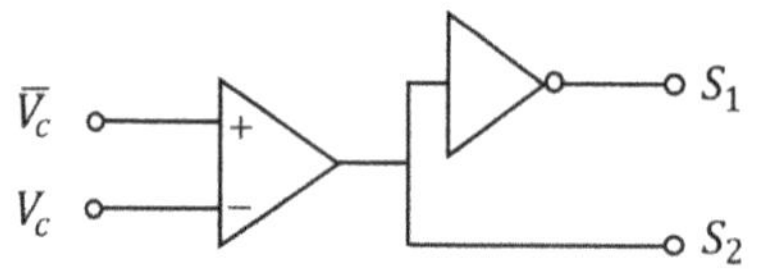

Figure 4. Control logic for IGBT group.

The total inserted voltage of the ESMs, V_M, in Figure 3, is determined by the number of ESMs inserted in each operational mode, N, and the capacitor voltage of the inserted ESM, according to (7). Additionally, capacitor balancing causes the ESM charging duty, V_{ci} at T_c, to be sorted in ascending order while the ESM discharging duty, V_{ci} at T_d, is sorted in descending order to ensure that the capacitor voltage of all ESMs is constantly charged and discharged.

$$V_M = \sum_{i=1}^{N} V_{ci} \tag{7}$$

At this time, the number of ESMs to be inserted N depends on N_{ctrl}, which is controlled by the output voltage, and a number of total modules n, which is determined by the specification of the NISD DC transformer. Therefore, the number of inserted ESMs N according to the charging and discharging cycle of the ESM is determined as (8). Also, as the ratio of input voltage to output voltage increases, the change of N, i.e., V_M, increases.

$$N = \begin{cases} n - N_{ctrl} & for\ T_c \\ N_{ctrl} & for\ T_d \end{cases} \tag{8}$$

Depending on the operational mode, the ESM is inserted to charge and discharge, and the output voltage is determined by M_1 and M_4. In order to effectively regulate the output voltage, the ESM must be inserted variably, and the control variable N_{ctrl} of the insertion number is controlled flexibly according to the output voltage. N_{ctrl} is calculated via the control diagram shown in Figure 5, and an additional insertion amount is calculated through the PI controller for error correction of the output voltage. The insertion amount is limited to a minimum of 1 and a maximum of $n - 1$ for the operation described in Figure 3. Since the number of modules is an integer, the amount of insertion is selected by converting to UInt (Unsigned Integer) type.

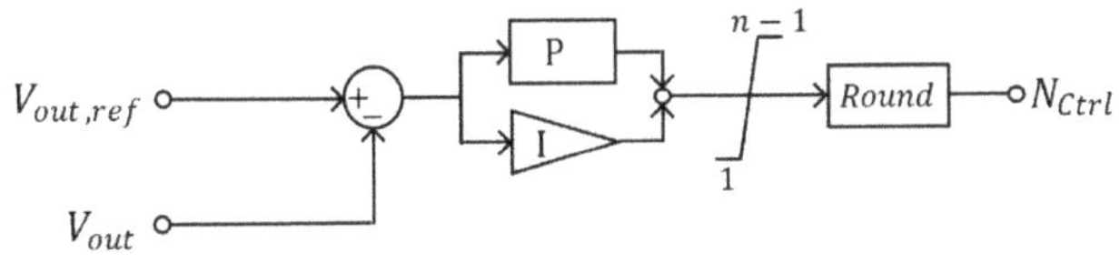

Figure 5. Control diagram for ESM insertion amount.

2.3.1. ESM Charging Duty (T_c)

Using (3) and (4), the voltage output during the ESM charging period at T_c is calculated from (9). At this time, the total inserted voltage of the ESMs, V_M between M_1 and M_3 is determined by (7) and (8).

$$V_{out} = V_{in} - \frac{t_2 - t_0}{t_1 - t_0} V_M \tag{9}$$

Meanwhile, the relationship between the output voltage V_{out} and the number N_{ctrl} of ESMs inserted while the ESMs are charging is shown in Figure 5. If the output voltage is higher than the reference voltage, additional modules are inserted to charge the ESMs and, simultaneously, the high electrical voltage difference between the input and output is reduced, i.e., if the output voltage is higher than the reference voltage, N_{ctrl} is reduced to increase the inserted number of ESMs, N. Alternatively, if the output voltage is lower than the reference voltage, N is reduced to control the output voltage.

2.3.2. ESM Discharging Duty (T_d)

The voltage output during the T_d discharge period of the ESM is obtained by utilizing (5) and (6), as demonstrated in (10). Similarly to T_c, (7) and (8) are applied to determine V_M in the M_4 to M_6 range.

$$V_{out} = \frac{t_5 - t_3}{t_4 - t_3} V_M - \frac{t_5 - t_4}{t_4 - t_3} V_{in} \tag{10}$$

The correlation between the output voltage V_{out} during the discharge of the ESMs and the control variable N_{ctrl} of the number of ESMs inserted is shown in Figure 5. The number of inserted ESMs, N, decreases when the output voltage is higher than the reference voltage, and, contrary to T_c, the number of inserted ESMs, N, decreases. As a result, the voltage source of the ESMs is reduced and the output voltage decreases. Conversely, as the output voltage is lower than the reference voltage, N_{ctrl} increases, which increases the number of ESMs inserted, and the output voltage is increased by the discharging ESMs.

3. Design Consideration of NISD DC Transformer

In this section, design considerations are provided for the implementation of the NISD DC transformer based on DCM. The energy conversion process is performed by multiple ESMs, including capacitors in the NISD DC transformer, and the overcurrent that can be produced by voltage unbalance during the conversion process is limited by inductors. Moreover, the capacitor on the output side reduces the ripple in output voltage.

Before analysis, Figure 2 demonstrates that when the NISD DC transformer operates by segments for $t_2 - t_0 > T_c$, i_L interrupts the ESM's capacitor discharge and load current in the T_d duty, which is not discharged in the T_c duty (i.e., $i_L > 0$). Similarly, when $t_5 - t_3 > T_d$, i_L interrupts the ESM's capacitor charge and load current in the T_c duty, which is not discharged in the T_d duty (i.e., $i_L < 0$). In both cases, the regular operation of the NISD DC transformer is interrupted, i.e., the theoretical maximum power transfer condition is $t_2 = t_3$ & $t_5 = t_6$, and when a load with more than the maximum power is connected, the output is reduced to protect itself from overcurrent. To ensure efficient operation, it is necessary for the maximum value I_{max} and minimum value I_{min} of the inductor current i_L to satisfy the condition $I_{max} = -I_{min}$, regarding the power loss I^2R in the conductor.

The values are determined by the switching frequency f_S of Group 1 ($S_{11}, \cdots, S_{1\alpha}$) and Group 2 ($S_{21}, \cdots, S_{2\alpha}$) IGBTs in Figure 1a and the switching frequency f_M of the two IGBTs (S_a, S_b) of the ESM in Figure 1b, which are influenced by the inductor current i_L at rated load conditions. Thus, in this section, the relationship among f_S, f_M, the inductor and capacitors and the quantity of input and output voltages, IGBTs and diodes, and ESMs in a NISD DC transformer will be discussed.

3.1. Voltage Stress on IGBTs and Diodes

IGBTs and diodes are arranged in groups of α and β ($\alpha, \beta \in \mathbb{N}$), respectively, to endure high voltages. IGBTs and diodes are controlled to operate, as shown in Figure 3, for voltage transformation.

IGBT Group 1 ($S_{11}, \cdots, S_{1\alpha}$) and Group 2 ($S_{21}, \cdots, S_{2\alpha}$), comprising α IGBTs, alternate in turning on/off to prevent short-circuiting of the NISD DC transformer. The voltage applied to the component when each IGBT group is operating is expressed as (11) and (12).

$$V_{S1} = V_{in} + V_{D2} - V_L - V_M \tag{11}$$

$$V_{S2} = V_{in} - V_{S1} - V_L \tag{12}$$

Diode groups 1 ($D_{11}, \cdots, D_{1\beta}$) and 2 ($D_{21}, \cdots, D_{2\beta}$) are composed of β diodes and perform opposite operations depending on the charging and discharging duty of the ESM. The voltage for each diode group is shown as (13), and, according to the equation, the reverse voltage during the reverse bias of the diode is mainly affected by V_{out}, and the voltage during the forward bias is operated to zero by the reverse voltage of the other diode group.

$$V_{D1} + V_{D2} = -V_{out} \tag{13}$$

In the design of the NISD DC transformer, the number of IGBTs and diodes is determined by $V_{in,ref}$ and $V_{out,ref}$, which are the reference of input and output voltages. From Figure 3, each group of components is organized and operated in series, so the number of IGBTs and diodes (α, β) for the rated voltage is determined as shown in (14) and (15).

When the rated collector-to-emitter voltage of an IGBT device is V_S, then

$$\alpha \geq \frac{V_{in,\,ref}}{V_S} \tag{14}$$

When the rated breakdown voltage of a diode device is V_D, then

$$\beta \geq \frac{V_{out,\,ref}}{|V_D|} \tag{15}$$

3.2. Current Calculations of IGBTs and Diodes

During operation, the IGBTs and diodes in each group are organized in series, as shown in Figure 3; thus, the current in each component within the same group is the same and is expressed as (16) and (17).

$$I_{S1} = I_{D1} = \begin{cases} I_L \ for \ T_c \\ 0 \ \ for \ T_d \end{cases} \tag{16}$$

$$I_{S2} = I_{D2} = \begin{cases} 0 \ \ for \ T_c \\ -I_L \ for \ T_d \end{cases} \tag{17}$$

3.3. Design on Inductor

The NISD DC transformer's inductor is installed to reduce the overcurrent that may occur when switching IGBTs' operation in accordance with the operating modes in relation to each duty ratio, and the inductance for the NISD DC transformer to operate can be calculated by the total voltage of the inserted ESM (7), which is calculated from the inductor voltage in (6) and the number of ESM inputs in this mode (8), and is calculated as shown in (18).

$$L = \frac{(n - N_{ctrl})N_{ctrl}V_{in}(nV_{out} - N_{ctrl}V_{in})}{n^3 I_{min} f_S (V_{out} - V_{in})} \tag{18}$$

3.4. Design on Capacitor

The capacitors in the NISD DC transformer can be categorized into two types: the internal capacitors of the ESM responsible for transforming energy and the capacitors aimed at reducing the output voltage's ripple.

The ESM's internal IGBTs (S_a, S_b), which are composed of n in series, regulate the ripples in ESM's voltage ΔV_M by operating at a switching frequency f_M. The capacitance C_M of the ESM's internal capacitor can be calculated using (19).

$$C_M = \frac{I_{max}(n - N_{ctrl})}{2n\Delta V_M f_M} \tag{19}$$

IGBT groups 1 ($S_{11}, \cdots, S_{1\alpha}$) and 2 ($S_{21}, \cdots, S_{2\alpha}$) operate at a switching frequency f_S, and their respective operational modes are determined. From this, the output capacitor current I_c and output voltage's ripple ΔV_{out} are determined, and the capacitance of the output capacitor C_{out} configured in parallel with the output terminal is expressed with (20).

$$C_{out} = \frac{I_c N_{ctrl}}{2n\Delta V_{out} f_S} \tag{20}$$

4. Case Studies

In this section, to verify the proposed NISD DC transformer, a DC transformer with an input voltage of 35 kV, an output voltage of 1.5 kV, and a rated power of 1 MW is configured with 35 ESMs. Moreover, the switching frequency is set to 5 kHz, and the ESM switching frequency is set to 50 kHz. In addition, the overcurrent limit inductor is 40 μH, the capacitance of each ESM is 20 mF, the output capacitor is 1.45 mF, and the simulation parameters are configured as shown in Table 1. The simulation is performed as shown in Figure 1 using PSCAD/EMTDC. Additionally, the performance of the control method for the capacitor's voltage is confirmed through seven cases.

Table 1. Parameters of the NISD DC transformer.

Symbol	Quantity	Value
V_{in}	Input voltage	35 kV
V_{out}	Output voltage	1.5 kV
P	Rated power	1 MW
n	Number of ESMs	35
f_S	Switching frequency	5 kHz
f_M	ESM switching frequency	50 kHz
L	Inductance	40 μH
C_{out}	ESM capacitance	20 mF
C_M	Output capacitance	1.45 mF

Figure 6a demonstrates that the output voltage is verified at 1.5 kV—with the output power being 1 MW and the output current at 0.667 kA—to achieve the rated power of 1 MW. Furthermore, ESM's insertion control variable N is regulated based on operational modes, with a range of 1 to 34, and the total injected voltage V_M of the ESM according to N is properly operated according to (7). The relationship between the current in the IGBTs and diodes and the inductor's current is also validated to perform the same as shown in

Figure 2—(16) and (17). The voltage applied to the IGBTs and diodes is also validated using a simulation, and the voltages in (11)–(13) are verified.

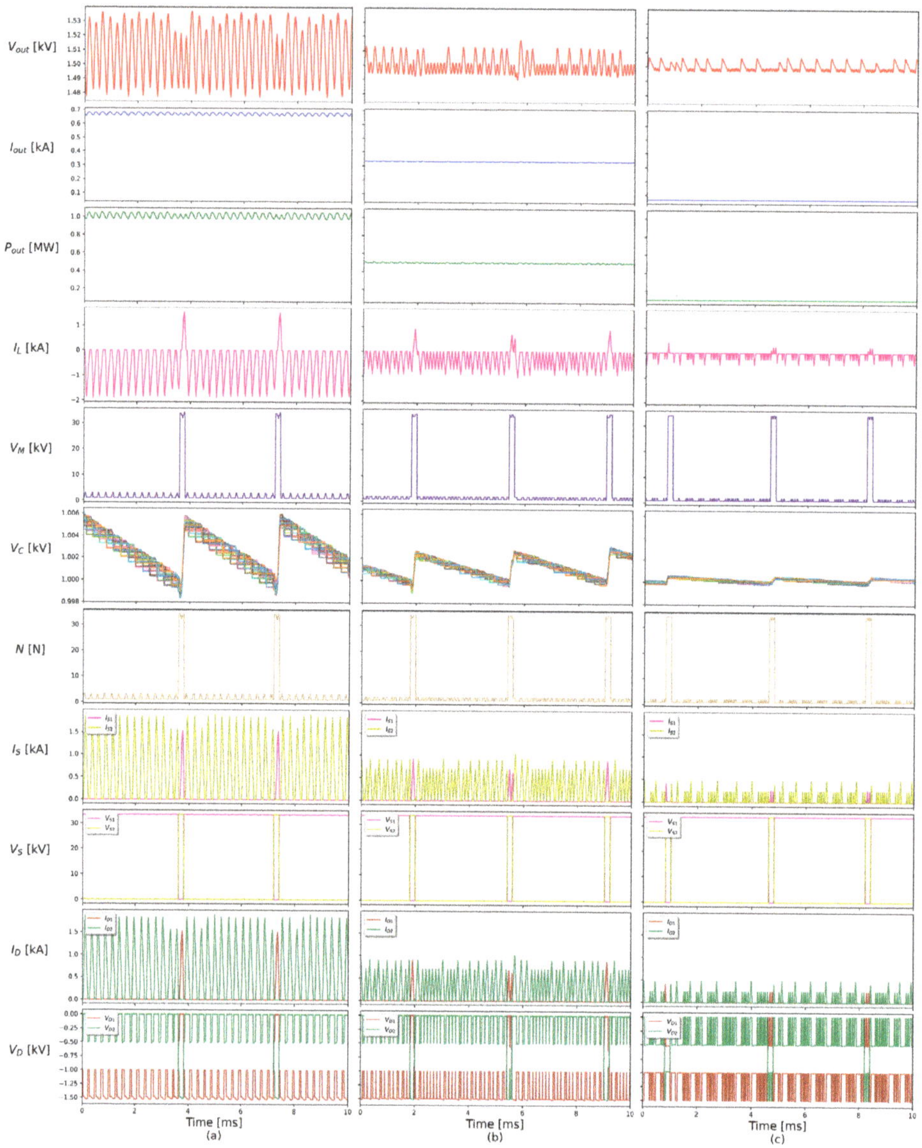

Figure 6. NISD DC transformer operation at resistive load of (**a**) 100%, (**b**) 50%, and (**c**) 10%.

4.1. Operations in Steady State

The NISD DC transformer is verified using constant resistive loads of 100%, 50%, and 10% (Figure 6a–c). The output voltage's ripples decrease by 3.904%, 1.44%, and 0.651%, respectively, as the load rate is reduced. Correspondingly, the output power's ripples decrease by 7.841%, 4.326%, and 1.304%, while the output voltage remains at 1.5 kV. The operational modes demonstrated consistency with the theoretical expectations, as the operating times of operational modes M_3 and M_6 increased with a decrease in load.

4.2. Operations with Disturbances

The simulation results, depicted in Figure 7a, illustrate normal output voltage control under the weak input voltage condition while operating under a 100% resistive load. in addition, the analysis of the figure indicates that the inductor current I_L is also influenced by the input at M_2, when compared to Figure 6a, at a weak input voltage.

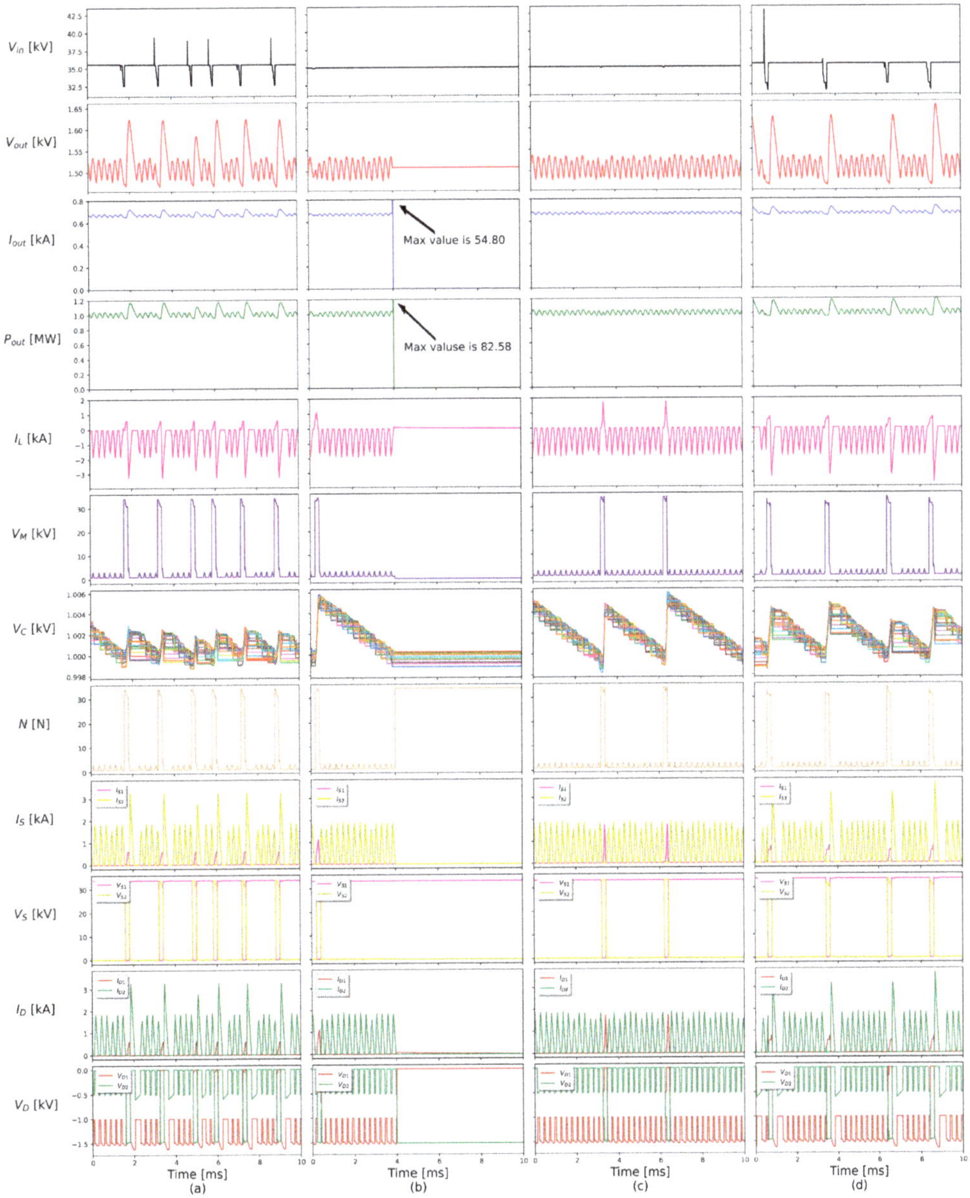

Figure 7. NISD DC transformer operation with (**a**) ripples in input voltage, (**b**) low-voltage side fault at 4 ms, (**c**) transition at 4 ms from 100% resistive to inductive load, and (**d**) transition at 4 ms from 100% resistive to inductive load during ripples in input voltage.

Next, in order to verify the operation and blocking ability in case of the LVDC side fault, a LVDC side fault was generated at 4 ms during an operation under 100% load, and the circuit breaker was opened 2 ms later to clear the fault (Figure 7b). Although, in this case, the fault current and power from the capacitor, C_{out}, increase to 54.80 kA and 82.58 MW, respectively. Such increases are a common problem for all inverters without a special protection scheme for an output capacitor. This might be solved by adding additional circuits, such as a damping resistor. Meanwhile, the rapid fault current reduction of the proposed converter based on the reduced capacitor size equates to a shorter discharge time with the same current compared to the other converters. At this time, all IGBTs were not damaged by immediately being opened. They cause the current and power to drop to zero and trigger sequential operations of the circuit breaker. Therefore, the proposed NISD DC transformer should be shortened. Inside the NISD DC transformer, the operation is similar to the second operational mode M_2, since all IGBTs are opened, and the voltage, V_{D2}, of diode group 2 is applied through the anti-parallel diode of IGBT group 2 and the diode of the ESM.

To evaluate the performance of the transformer during the rapid load changes of the power system, an inductance of 0.1 H was transitioned at 4 ms while operating at 100% resistive load rate (Figure 7c). This representatively reflects the aggregated effect caused by independent changes of various loads located through different feeders, and the system is stable.

Finally, the inductance was transitioned at 4 ms with high-impedance input voltage, i.e., the conditions in Figure 7a,c were considered simultaneously. Therefore, a weak input voltage causes the output current to ripple up to 8.077% due to the discharge of the capacitor, and the output power to ripple up to 15.728%. Although the inductor current is affected, it operates between 0.798 kA and −2.819 kA. Thus, it does not cause a problem for IGBT group 1 ($S_{11}, \cdots, S_{1\alpha}$) and 2 ($S_{21}, \cdots, S_{2\alpha}$) and diode group 1 ($D_{11}, \cdots, D_{1\beta}$) and 2 ($D_{21}, \cdots, D_{2\beta}$). Also, the increase in the ripple of output voltage is negligible. As shown in Figure 7d, the NISD DC transformer operates reliably, even in harsh conditions.

5. Conclusions

This article proposed a non-isolated step-down (NISD) DC transformer based on DCM for the MVDC-LVDC distribution system. To verify the performance of the proposed NISD DC transformer, normal operation was confirmed with loads of 100%, 50%, and 10%. Additionally, abnormal situations such as inductive load, LVDC side fault, and instability input voltage were assumed and simulated using a power system computer-aided design and electromagnetic transient simulation engine including DC (PSCAD/EMTDC). The proposed NISD DC transformer successfully operated for the purpose of energy transformation.

This study verified the performance of energy transformation from MVDC to LVDC, i.e., the MVDC-LVDC distribution system. However, further studies are required to apply the proposed NISD DC transformer to an HVDC-MVDC system or step-up version of the proposed transformer, even a bidirectional one.

Author Contributions: Conceptualization, S.H.L.; methodology, S.H.L. and D.C.; software, M.K.; validation, S.H.L. and M.K.; formal analysis, S.H.L. and M.K.; investigation, S.H.L. and M.K.; resources, S.H.L. and M.K.; data curation, S.H.L. and M.K.; writing—original draft preparation, M.K.; writing—review and editing, S.H.L.; visualization, M.K.; supervision, S.H.L. and D.C.; project administration, S.H.L.; funding acquisition, S.H.L. and D.C. All authors have read and agreed to the published version of the manuscript.

Funding: This work was supported in part by the National Research Foundation of Korea (NRF) funded by the Korean Government (MSIT) under grant 2022R1A2C2006688, and in part by the Basic Science Research Program through the National Research Foundation of Korea (NRF) funded by the Ministry of Education (2022R1C1C1005975).

Data Availability Statement: The datasets presented in this article are not readily available because the data are part of an ongoing study.

Conflicts of Interest: All of the authors declare that they have no known competing financial interests or personal relationships that could have appeared to influence the work reported in this paper.

References

1. Enerdata. *Global Energy Trends 2023*; Enerdata: Grenoble, France, 2023.
2. Suryadevara, R.; Parsa, L. Full-Bridge ZCS-Converter-Based High-Gain Modular DC-DC Converter for PV Integration with Medium-Voltage DC Grids. *IEEE Trans. Energy Convers.* **2019**, *34*, 302–312. [CrossRef]
3. Lu, Z.; Li, C.; Zhu, A.; Luo, H.; Li, C.; Li, W.; He, X. Medium Voltage Soft-Switching DC/DC Converter with Series-Connected SiC MOSFETs. *IEEE Trans. Power Electron.* **2021**, *36*, 1451–1462. [CrossRef]
4. Jin, H.; Chen, W.; Hou, K.; Shao, S.; Shu, L.; Li, R. A Sharing-Branch Modular Multilevel DC Transformer with Wide Voltage Range Regulation for DC Distribution Grids. *IEEE Trans. Power Electron.* **2022**, *37*, 5714–5730. [CrossRef]
5. *IEEE Standard 1709-2018*; IEEE Recommended Practice for 1 kV to 35 kV Medium-Voltage DC Power Systems on Ships. IEEE: New York, NY, USA, 2018.
6. Yao, J.; Chen, W.; Xue, C.; Yuan, Y.; Wang, T. An ISOP Hybrid DC Transformer Combining Multiple SRCs and DAB Converters to Interconnect MVDC and LVDC Distribution Networks. *IEEE Trans. Power Electron.* **2020**, *35*, 11442–11452. [CrossRef]
7. Reddy, P.L.S.K.; Obulesu, Y.P. Design and Development of a New Transformerless Multi-port DC–DC Boost Converter. *J. Electr. Eng. Technol.* **2023**, *18*, 1013–1028. [CrossRef]
8. Zhao, B.; Song, Q.; Li, J.; Sun, Q.; Liu, W. Full-Process Operation, Control, and Experiments of Modular High-Frequency-Link DC Transformer Based on Dual Active Bridge for Flexible MVDC Distribution: A Practical Tutorial. *IEEE Trans. Power Electron.* **2017**, *32*, 6751–6766. [CrossRef]
9. Sheng, J.; Chen, C.; Lu, R.; Li, C.; Xiang, X.; Li, W.; He, X. Control Optimization of Modular Multilevel Resonant DC Converters for Wide-Input-Range MVdc to LVdc Applications. *IEEE Trans. Power Electron.* **2022**, *37*, 5284–5298. [CrossRef]
10. Ferreira, J.A. The multilevel modular DC converter. *IEEE Trans. Power Electron.* **2013**, *28*, 4460–4465. [CrossRef]
11. Jovcic, D. Bidirectional, high-power DC transformer. *IEEE Trans. Power Deliv.* **2009**, *24*, 2276–2283. [CrossRef]
12. Li, W.; He, X. Review of Nonisolated High-Step-Up DC/DC Converters in Photovoltaic Grid-Connected Applications. *IEEE Trans. Ind. Electron.* **2011**, *58*, 1239–1250. [CrossRef]
13. Stieneker, M.; De Doncker, R.W. Medium-voltage DC distribution grids in urban areas. In Proceedings of the 2016 IEEE 7th International Symposium on Power Electronics for Distributed Generation Systems (PEDG), Vancouver, BC, Canada, 27–30 June 2016; IEEE: Vancouver, BC, Canada, 2016; pp. 1–7. [CrossRef]
14. Luth, T.; Merlin, M.M.C.; Green, T.C.; Hassan, F.; Barker, C.D. High-Frequency Operation of a DC/AC/DC System for HVDC Applications. *IEEE Trans. Power Electron.* **2014**, *29*, 4107–4115. [CrossRef]
15. Wang, X.; Peng, Y.; Chai, J.; Xia, Y.; Wei, W.; Yu, M. An Ideal DC Transformer for Active DC Distribution Networks Based on Constant-Transformation-Ratio DABC. *IEEE Trans. Power Electron.* **2020**, *35*, 2170–2183. [CrossRef]
16. Zhao, B.; Song, Q.; Liu, W.; Sun, Y. Overview of Dual-Active-Bridge Isolated Bidirectional DC–DC Converter for High-Frequency-Link Power-Conversion System. *IEEE Trans. Power Electron.* **2014**, *29*, 4091–4106. [CrossRef]
17. Alhurayyis, I.; Elkhateb, A.; Morrow, J. Isolated and Nonisolated DC-to-DC Converters for Medium-Voltage DC Networks: A Review. *IEEE J. Emerg. Sel. Top. Power Electron.* **2021**, *9*, 7486–7500. [CrossRef]
18. Xiang, X.; Gu, Y.; Qiao, Y.; Zhang, X.; Li, W.; He, X.; Green, T.C. Resonant Modular Multilevel DC–DC Converters for Both High and Low Step-Ratio Connections in MVDC Distribution Systems. *IEEE Trans. Power Electron.* **2021**, *36*, 7625–7640. [CrossRef]
19. Zhao, B.; Li, X.; Wei, Y.; An, F.; Sun, J.; Liu, G.; Zhao, Y. Modular Hybrid-Full-Bridge DC Transformer with Full-Process Matching Switching Strategy for MVdc Power Distribution Application. *IEEE Trans. Ind. Electron.* **2020**, *67*, 3317–3328. [CrossRef]
20. Du, S.; Wu, B.; Tian, K.; Xu, D.; Zargari, N.R. A Novel Medium-Voltage Modular Multilevel DC–DC Converter. *IEEE Trans. Ind. Electron.* **2016**, *63*, 7939–7949. [CrossRef]
21. Gray, P.A.; Lehn, P.W.; Yakop, N. A Modular Multilevel DC–DC Converter with Flying Capacitor Converter Like Properties. *IEEE Trans. Ind. Electron.* **2022**, *69*, 6774–6783. [CrossRef]
22. Xing, D.; Hu, B.; Li, X.; Cheng, Q.; Wang, J.; Atcitty, S. Comparison of Modular Multilevel Converter Based High Voltage-Step-Down Ratio DC/DC Converters for Energy Storage System Applications. *IEEE J. Emerg. Sel. Top. Power Electron.* **2023**. [CrossRef]
23. Falin, J.; Aguilar, A. Maximize power density with three-level buck-switching charges. *Analog. Des. J. Q* **2021**, 1–8.
24. Meynard, T.A.; Foch, H. Multi-level conversion: High voltage choppers and voltage-source inverters. In Proceedings of the PESC '92 Record, 23rd Annual IEEE Power Electronics Specialists Conference, Toledo, Spain, 29 June–31 July 1992; 1992; Volume 1, pp. 397–403. [CrossRef]
25. Stupar, A.; McRae, T.; Vukadinović, N.; Prodić, A.; Taylor, J.A. Multi-Objective Optimization of Multi-Level DC–DC Converters Using Geometric Programming. *IEEE Trans. Power Electron.* **2019**, *34*, 11912–11939. [CrossRef]
26. Ye, Z.; Lei, Y.; Liao, Z.; Pilawa-Podgurski, R.C.N. Investigation of Capacitor Voltage Balancing in Practical Implementations of Flying Capacitor Multilevel Converters. *IEEE Trans. Power Electron.* **2022**, *37*, 2921–2935. [CrossRef]
27. Villar, G.; Alarcon, E. Monolithic Integration of a 3-Level DCM-Operated Low-Floating-Capacitor Buck Converter for DC-DC Step-Down Conversion in Standard CMOS. In Proceedings of the 2008 IEEE Power Electronics Specialists Conference, Rhodes, Greece, 15–19 June 2008. [CrossRef]

28. Denniston, N.; Massoud, A.M.; Ahmed, S.; Enjeti, P.N. Multiple-Module High-Gain High-Voltage DC–DC Transformers for Offshore Wind Energy Systems. *IEEE Trans. Ind. Electron.* **2011**, *58*, 1877–1886. [CrossRef]
29. Athab, H.; Yazdani, A.; Wu, B. A transformerless DC-DC converter with large voltage ratio for MV DC grids. *IEEE Trans. Power Deliv.* **2014**, *29*, 1877–1885. [CrossRef]
30. Abbasi, M.; Emamalipour, R.; Cheema, M.A.M.; Lam, J. A Constant-Frequency High-Voltage Gain Resonant Converter Module with Semiactive Phase-Shifted Voltage Multiplier for MVdc Distribution. *IEEE J. Emerg. Sel. Top. Power Electron.* **2022**, *10*, 3603–3616. [CrossRef]
31. Zheng, L.; Kandula, R.P.; Divan, D. Current-Source Solid-State DC Transformer Integrating LVDC Microgrid, Energy Storage, and Renewable Energy into MVDC Grid. *IEEE Trans. Power Electron.* **2022**, *37*, 1044–1058. [CrossRef]
32. Li, L.; Jiang, Z.; Jiang, Y. A SEPIC Flyback DC-DC Converter with GaN Switching Device. *J. Electr. Eng. Technol.* **2023**, *18*, 2759–2769. [CrossRef]

energies

MDPI

Article

An Optimized Switching Strategy Based on Gate Drivers with Variable Voltage to Improve the Switching Performance of SiC MOSFET Modules

Jixiang Tan * and Zhongfu Zhou

Department of Electronic and Electrical Engineering, Faculty of Science and Engineering, Swansea University Bay Campus, Swansea SA1 8EN, UK; z.zhou@swansea.ac.uk
* Correspondence: 979921@swansea.ac.uk

Abstract: This paper proposes an optimized switching strategy (OSS) based on a silicon carbide (SiC) MOSFET gate driver with variable voltage, which allows simultaneous variations in several different parameters to optimize the switching performance of semiconductor devices. As a relatively new device, the SiC MOSFET shines in the field of high power density and high-frequency switching; it has become a popular solution for electric vehicles and renewable energy conversion systems. However, the increase in voltage and current slope caused by high switching speeds inevitably increases the overshoot and oscillation in a circuit and can even generate additional losses. The principle of this new control strategy is to change the voltage and current in the turn-on and turn-off stages by changing the gate driver's voltage. That is, we reduced the drive's voltage after a certain time delay and maintained it for a period of time, thus directly controlling the slopes of di/dt and dv/dt. This study focused on the optimization of the SiC MOSFET by changing the time delay preceding the decrease in the voltage of the gate driver, analyzing and calculating the optimal time delay before the decrease in the voltage of the gate driver, and verifying the findings using LTspice simulation software. The simulated results were compared and analyzed with hard-switching strategies. The results showed that the proposed OSS can improve the switching performance of SiC MOSFETs.

Keywords: optimal-switching strategy; gate driver circuit; silicon carbide MOSFET; voltage and current overshoot; LTspice

1. Introduction

Silicon carbide (SiC) MOSFETs offer a range of advantages over silicon-based switches, including faster switching, higher efficiency, higher operating voltages, and higher temperatures, enabling smaller and lighter designs [1,2]. Thus, significant progress has been made in the power semiconductor industry. These advantages allow SiC MOSFETs to be applied in a range of automotive and industrial applications. However, the excessively fast switching speed of SiC MOSFETs causes high values of di/dt and dv/dt [3]. At the same time, due to the existence of parasitic inductance and parasitic resistance in circuits, SiC MOSFETs are prone to overshoots and oscillations in current and voltage during switching, thus generating additional switching losses. These losses can even cause device damage. Therefore, reducing or even eliminating the overshoot, oscillation, and electromagnetic interference (EMI) of silicon carbide MOSFETs during switching is a priority in improving their working efficiency [4].

Current and voltage overshoots are the biggest obstacles in SiC semiconductor applications. Stray inductance and parasitic capacitance are the main contributors to current and voltage overshoots [5,6]. When a SiC MOSFET works at a high frequency, the switching speed becomes too fast and the di/dt and dv/dt slopes increase. Thus, current and voltage overshoots are amplified due to stray inductance and parasitic capacitance [7]. When the voltage and current overshoots exceed the breakdown voltage of the SiC MOSFET and the maximum recovery current of the freewheeling diode, the SiC is destroyed. However, the parasitic elements in the actual circuit cannot be eliminated [8–11]; so, the overshoot can only be eliminated by other means.

Solving the problems of overshoots and oscillations is key to improving the efficiency of SiC MOSFETs.

In the traditional approach, slowing the switching speed in the case of high gate resistance [12] can alleviate the SiC MOSFET's overshoots and oscillations. However, this solution means the power losses become greater with longer switching times [6]. Adding an RC snubber circuit is also a common method [13,14] to eliminate overshoot and oscillation problems. Although highly efficient snubber circuits have been proposed, snubber circuits affect the overall efficiency of the system. In particular, extra devices increase the power loss of the overall circuit. Active gate driver (AGD) solutions have been developed to increase the efficiency of power devices. The main advantages of these gate drivers are reduced oscillations and overshoots. However, the extra components not only increase the volume of the SiC system but also increase the circuit's power loss [15–21]. In [21], a new active gate driver was proposed that could effectively suppress overshoots and oscillation and reduce losses without compromising EMI. The main strategy of the proposed AGD was to reduce the current and voltage slope by reducing the gate driver's voltage. However, its main disadvantages were that the process of calculating the parameters of the transformer was complicated, an additional printed circuit board was required, and the implementation of the circuit was complicated and expensive. As such, configurable digital gate drivers (DGDs) are the latest technological development in this sphere [22]. The strengths of DGDs are their programmability, support for a wide variety of power devices, and ease of use [23]. A key element of this technology is the ability to configure the turn-on and turn-off processes, which provides a series of steps to control the voltage level at specific times [24–28]. This allows designers to digitally configure the turn-on and turn-off curve through software without requiring changes to the hardware.

In this paper, an optimized switching strategy based on gate drivers with variable voltage (OSS) was proposed to improve the switching performance of SiC MOSFETs. This switching strategy was based on and optimized for an AGD in [21] and a DGD in [22]. There is no doubt that the AGD performed well in [21]. However, the gate driver circuits were too complex and too large. The ease of operation of the DGD in [22] is very attractive. Therefore, this article aimed to combine the advantages of the voltage gate drivers of AGDs and DGDs in waveforms to further optimize the algorithm and control and reduce overshoots and oscillations in SiC MOSFETs through an OSS. Finally, we used the LTspice simulation software to verify the feasibility of our switching strategy. Compared with AGDs and DGDs, an OSS can control more variables to achieve more precise and optimal circuit control. Due to space limitations, this article focused on the impact of delay time on switching performance and lists the formula to calculate the delay time for a given calculated decrease in voltage. At the same time, the decrease in the voltage and the time of this decrease were also considered. The development potential of the OSS is significant. Compared with other switching strategies, which can only change one or two variables, the OSS can control three variables simultaneously, namely, the delay time before a decrease in voltage, the decrease in voltage itself, and the time duration of the decrease in voltage. More variables mean better control performance. However, due to various limitations, this paper only considered the influence of the delay time before a decrease in voltage on switch performance and analyzed the relationship between voltage drop and switch performance. In our next work, we will produce more simulations and experimental results to prove the advanced nature of the OSS.

This paper is organized as follows. Sections 2 and 3 present the control principle, circuit model, and related formulations of the proposed switching strategy. In Section 4, the simulation setup and simulation results are introduced and analyzed. Finally, in Section 5, the conclusions and a discussion of this simulation are given.

2. Operation Principle of the OSS

The working principle of the OSS is to improve the switching performance of the SiC MOSFET by controlling the voltage drop delay time, voltage drop, and voltage drop time. Compared with traditional AGDs and DGDs, the advantage of an OSS is that it has more control parameters, which means that the semiconductors can be optimally controlled. In this section, the general schematic circuit and working principle of the proposed OSS will be introduced in detail. Figure 1 shows a circuit schematic of the SiC MOSFET and Figure 2 shows a timing chart of the SiC MOSFET with an optimized switching strategy.

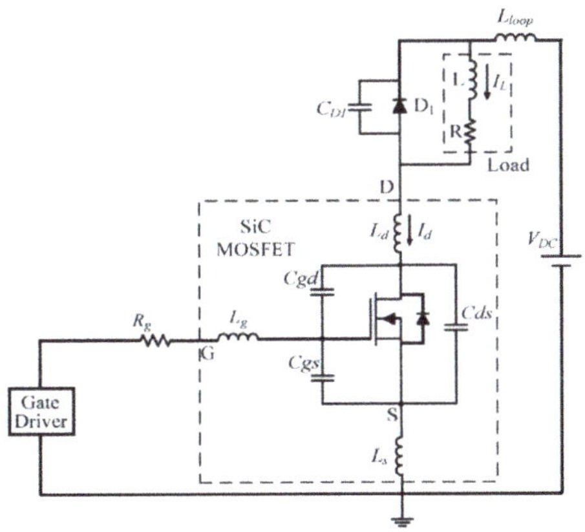

Figure 1. Circuit schematic of the SiC MOSFET under test.

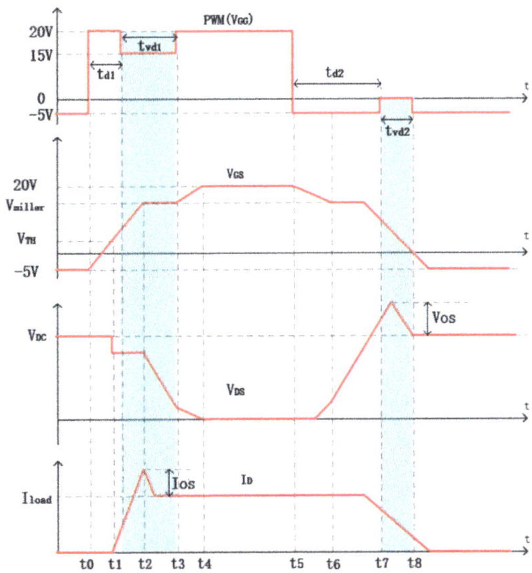

Figure 2. Timing chart of the SiC MOSFET with the OSS.

2.1. Operation Modes

As shown in Figure 1, the OSS circuit was mainly composed of three parts: the gate drive circuit, silicon carbide MOSFET, and DC bus power supply. The model of the SiC MOSFET, including junction capacitances (C_{gd}, C_{gs}, C_{ds}) and module intrinsic parasitic inductances (Ls, Lg, Ld), is also depicted in Figure 1.

In [21], the gate drive voltage used by the AGD was 20 V/−5 V; also, 15 V and 0 V were used as dropped voltages to optimize the switching performance of the SiC MOSFET. Because it used a switching bridge, it could generate four gate drive voltages (i.e., 20 V, 15 V, 0, and −5 V) from two drive power supplies (20 V and 5 V) through different combinations. In this design, the gate drive voltage waveform design of the AGD was still used. The purpose was to control variables and only explore the influence of the voltage drop delay time (t_d) on the performance of SiC MOSFETs.

2.2. Operation Principle

Figure 2 shows the timing chart of the SiC MOSFET with an optimized switching strategy. The whole process could be split into two steps: turn on and turn off. The blue interval represents the gate driver voltage in the stage of the voltage drop delay time.

2.2.1. Turn-On Stage

When the pulse width modulation (PWM) signal changed from low to high at t_0, a high value gate drive voltage V_{GG} of 20 V was applied to the SiC MOSFET. Then, a high gate current was generated to charge the input capacitance $C_{iss} = C_{gs} + C_{gd}$ with R_g, which meant the V_{GS} started to increase. When the V_{GS} reached the threshold voltage V_{TH} at t_1, the I_D began to conduct. After a time delay t_{d1}, the lower V_{GG} was activated and a lower gate current was generated because the V_{GG} was reduced to 15 V.

In this interval, I_D continued to rise and a current peak value I_{OS} appeared due to the freewheeling diode effect. The V_{GS} reached the Miller plateau voltage and stopped rising. At the same time, the V_{DS} started to drop. After the voltage fall time t_{vd1}, the V_{GS} came out of the Miller plateau and continued to rise. Finally, the SiC MOSFET was fully turned on at t_4. From [21], the current slope of I_D and the voltage slope of the V_{DS} during the turn-on stage could be expressed as:

$$\frac{dI_D}{dt} = \frac{V_{GG} - V_{TH} - I_D/g_m}{(C_{iss} \cdot R_{on})/g_m + L_s} \tag{1}$$

$$\frac{dV_{DS}}{dt} = -\frac{V_{GG} - V_{Miller}}{C_{GD} \cdot R_{on}} \tag{2}$$

where g_m is the transconductance of the SiC MOSFET and C_{gd} is the gate-drain capacitance of the SiC MOSFET.

2.2.2. Turn-Off Stage

When the signal from the voltage source changed from high to low at t_5, a negative voltage (i.e., −5 V) was generated to discharge the C_{iss} with R_g. The V_{GS} started to drop from 20 V until it reached the Miller voltage at t_6. The V_{DS} rose rapidly and caused a voltage overshoot V_{OS} due to stray inductance. After a delay time t_{d2}, the V_{GG} increased to 0 V and remained there until I_D was fully turned off at t_8. After t_8, the gate voltage stabilized at −5 V to ensure that the SiC MOSFET was always in the off state. The voltage slope could be calculated as:

$$\frac{dI_D}{dt} = g_m \cdot \frac{V_{TH} + (I_D/g_m) - V_{GG}}{(C_{iss} \cdot R_{off})/g_m + L_s} \tag{3}$$

$$\frac{dV_{DS}}{dt} = \frac{V_{GG} - V_{Miller}}{C_{GD} \cdot R_{off}} \tag{4}$$

According to (1)–(4), changing the instantaneous slope of current and voltage could be achieved by changing the V_{GG}. In the turn-on stage, the change was reducing the gate diver voltage while in the turn-off stage, the change was increasing the gate diver voltage.

3. Calculation for Voltage Drop Delay Time

An OSS can change the switching performance through unlimited control of the voltage drop delay time, voltage drop, and voltage drop time, aiming to obtain the best data for the switching performance through countless permutations and combinations. However, this is difficult to carry out. In fact, the best data can be derived by equation. The main research direction of this paper was to explore the influence of the voltage drop delay time on the switching performance and to explore the influence of different voltage drops and different voltage drop times on the switching performance. Thus, the content of this chapter focuses on deriving the best voltage drop delay time by equation.

3.1. Equivalent Slope for Current and Voltage

The slope of the current and voltage is an important factor affecting the overshoot and oscillation during the switching process of the SiC MOSFET. It can be seen from Figure 2 that the I_{OS} in the turn-on stage and the V_{OS} in the turn-off stage were obvious and large. However, the slope of the voltage and current in the SiC MOSFET with the OSS method was not constant because of the variable gate drive voltage; so, the equivalent slope was proposed and applied in the calculation, with the aim of writing the voltage drop delay times, t_{d1} and t_{d2}, into the equation.

In Figure 3, when the V_{GG} was 20 V, the slope of the current was dI_D/dt_1; when the V_{GG} was 15 V, the slope of the current was dI_D/dt_2. Similarly, dV_{DC}/dt_1 and dV_{DC}/dt_2 represented the value of the voltage slope under the V_{GG} of −5 and 0 V. In addition, ton represented the increase in the current I_D and toff represented the increase in the voltage V_{DS}; t_{on1} and t_{off1} were the rising times under dI_D/dt_1 and dV_{DC}/dt_1, respectively. Moreover, dI_D/dt represented the actual current slope with OSS and dV_{DC}/dt was the current slope with OSS, in actuality. Additionally, dI_D/dt_{eq} was the equivalent slope at the turn-on stage and dV_{DC}/dt_{eq} represented the equivalent slope at the turn-off stage; t_{on2} and t_{off2} were the actual running times for the turn-on stage and turn-off stage. In [21], the slope of the current was used in both the turn-on and turn-off stages because it was convenient for the subsequent calculations. In this paper, in order to fit reality, the voltage slope was used in the turn-off stage. According to the equivalent principle shown in Figure 3, the following equation could be obtained:

$$I_L = |dI_D/dt_1|\cdot t_{on} + |dI_D/dt_1|\cdot(t_{d1} - t_{on}) + |dI_D/dt_2|\cdot(t_{on2} - t_{d1}) = |dI_D/dt_{eq}|\cdot t_{on2} \quad (5)$$

and the equivalent rising time t_{on2} of the current can be expressed as:

$$t_{on2} = (I_L - (|dI_D/dt_1| - |dI_D/dt_2|)\cdot t_{d1})/|dI_D/dt_2| \quad (6)$$

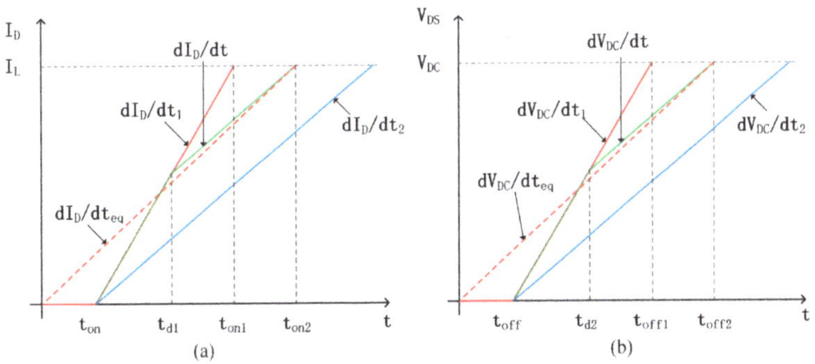

Figure 3. Schematic diagram of the equivalent slope. (**a**) Current slope on the turn-on stage. (**b**) Voltage slope on the turn-off stage.

Thus, according to (5) and (6), the value of the equivalent current slope at the turn-on stage can be expressed as:

$$|dI_D/dt_{eq}| = I_L \cdot |dI_D/dt_2| / (I_L - (|dI_D/dt_1| - |dI_D/dt_2|) \cdot t_{d1}) \quad (7)$$

Similarly, at the turn-off stage, the equations of t_{off2} and voltage slope are:

$$V_{DC} = |dV_{DS}/dt_1| \cdot t_{off} + |dV_{DS}/dt_1| \cdot (t_{d2} - t_{off}) + |dV_{DS}/dt_2| \cdot (t_{off2} - t_{d2}) = |dI_D/dt_{eq}| \cdot t_{off2} \quad (8)$$

$$t_{off2} = (V_{DC} - (|dV_{DS}/dt_1| - |dV_{DS}/dt_2|) \cdot t_{d2}) / |dV_{DS}/dt_2| \quad (9)$$

$$|dV_{DS}/dt_{eq}| = V_{DC} \cdot |dV_{DS}/dt_2| / V_{DC} - (|dV_{DS}/dt_1| - |dV_{DS}/dt_2|) \cdot t_{d2} \quad (10)$$

3.2. Power Losses in the Turn-On and Turn-Off Stages

The turn-on and turn-off losses of the power device could be calculated by (11) and (12), which were derived in []:

$$E_{ON} = E_{on,dI_D/dt} + E_{on,dV_{DS}/dt} + E_{Irr} + E_{Ls} \quad (11)$$

$$E_{OFF} = E_{off,dI_D/dt} + E_{off,dV_{DS}/dt} + E_{Ls} \quad (12)$$

where E_{Ls} is the energy generated by the current passing through L_S and E_{Irr} is the energy loss of the device generated by the reverse recovery effect, which can be calculated by (14) [29]:

$$E_{Ls} = \frac{1}{2} \cdot L_S \cdot {I_{loop}}^2 \quad (13)$$

$$E_{Irr} = \left(I_L \cdot \sqrt{Q_{rr}/|dI_D/dt|} + Q_{rr}\right) \cdot V_{DC} \cdot (1 - \sigma) \quad (14)$$

where L_{loop} is the stray inductance in the power loop, Q_{rr} is the reverse recovery charge, and σ is the overshoot ratio, which can be defined as:

$$\sigma = \frac{V_{OS}}{V_{DC}} = \frac{L_{loop} \cdot |dI_D/dt|}{V_{DC}} \quad (15)$$

During the turn-on transient processes, the energy losses during the current and voltage changes can be expressed as:

$$E_{on,dI_D/dt} = \frac{1}{2} \cdot I_L \cdot \left(V_{DC} - L_{loop} \cdot |dI_D/dt|\right) \cdot \frac{I_L}{|dI_D/dt|} \quad (16)$$

$$E_{on,dV_{DS}/dt} = \frac{1}{2} \cdot I_L \cdot V_{DC} \cdot \frac{V_{DC}}{|dV_{DS}/dt|} \cdot (1 - \sigma)^2 \quad (17)$$

Therefore, according to Formulas (11) and (13)–(17), the energy loss during the turn-on stage can be expressed as:

$$E_{ON} = \frac{V_{DC} \cdot {I_L}^2}{2} \cdot \left(\frac{1 + (1 - \sigma)^2}{|dI_D/dt|}\right) + \left(I_L \cdot \sqrt{\frac{Q_{rr}}{|dI_D/dt|}} + Q_{rr}\right) \cdot V_{DC} \cdot (1 - \sigma) - \frac{1}{2} \cdot L_S \cdot {I_{loop}}^2 \quad (18)$$

Similarly, according to (12), the energy loss can be expressed as:

$$E_{OFF} = \frac{I_L \cdot {V_{DC}}^2}{2} \cdot \frac{1 + (1 + \sigma)^2}{|dV_{DS}/dt|} + \frac{1}{2} \cdot L_S \cdot {I_{loop}}^2 \quad (19)$$

3.3. Calculation of the Voltage Drop Delay Time

In order to simplify the calculation model, it was assumed that the change of power consumption was only related to overshoot and was positively correlated. Therefore, during the turn-on stage of the SiC MOSFET, the change in power consumption is positively correlated with the change in current overshoot [21]. Similarly, in the turn-off phase, the change of power consumption is positively correlated with the change of voltage overshoot. The formula can be expressed as:

$$\frac{|E_{ON}\,(t_{d1}) - E_{ON}\,(0)|}{|E_{ON}\,(t_{on2}) - E_{ON}\,(0)|} = \frac{|I_{OS}(t_{d1}) - I_{OS}(0)|}{|I_{OS}(t_{on2}) - I_{OS}(0)|} \tag{20}$$

$$\frac{|E_{OFF}\,(t_{d2}) - E_{OFF}\,(0)|}{|E_{OFF}\,(t_{off2}) - E_{OFF}\,(0)|} = \frac{|V_{OS}(t_{d2}) - V_{OS}(0)|}{|V_{OS}(t_{off2}) - V_{OS}(0)|} \tag{21}$$

In the turn-on phase, the most important factor affecting power consumption is the overshoot of the current; so, Equation (21) can be abbreviated as:

$$E_{ON} = \frac{{I_L}^2 \cdot V_{DC}}{2} \cdot \frac{1-\sigma}{|dI_D/dt|} \tag{22}$$

Because the main cause of current overshoot is the reverse recovery effect [21], the current overshoot can be expressed as:

$$I_{OS} = I_{rr} = \sqrt{Q_{rr} \cdot |dI_D/dt|} \tag{23}$$

According to (25) and (26), (11) can be expressed as:

$$\frac{|dI_D/dt_{eq}|}{|dI_D/dt_1|} = \frac{\sqrt{Q_{rr} \cdot |dI_D/dt_1|}}{\sqrt{Q_{rr} \cdot |dI_D/dt_{eq}|}} \tag{24}$$

Similarly, in the turn-off stage, the formula can be obtained:

$$E_{OFF} = \frac{{V_{DC}}^2 \cdot I_L}{2} \cdot \frac{1+\sigma}{|dV_{DS}/dt|} \tag{25}$$

$$\frac{|dV_{DS}/dt_{eq}|}{|dV_{DS}/dt_1|} = \frac{L_{loop} \cdot |dV_{DS}/dt_1|}{L_{loop} \cdot |dV_{DS}/dt_{eq}|} \tag{26}$$

According to (1)–(4), (7), and (24), an equation expressed by t_{d1} could be obtained. Similarly, according to (1)–(4), (10), and (26), an equation expressed by t_{d2} could be obtained. Therefore, the optimal voltage drop delay times, t_{d1} and t_{d2}, could be calculated.

4. Simulation Verification

In order to evaluate the performance of the proposed OSS, a double-pulse simulation test of the SiC MOSFET was performed using the software LTspice. The power device used in the test was CREE's 1.2 KV/115 A SiC MOSFET module (C3M0016120D). In the simulation, the tested module was first modeled in LTspice to ensure that the simulation results would be closer to the experimental data and then the circuit was built and simulated. The detailed parameters of the tested modules in the experiment are shown in Table 1.

The comparison of the OSS's performance under different time delays was simulated to verify whether the calculated time delay was the best. To further compare the performance of the OSS, it was compared with hard switching (HS) and a digital gate driver (DGD) from Agileswitch.

Table 1. Parameters of the SiC MOSFET C3M0016120D (TCJ = 25 °C).

Symbol	Parameter	Value	Unit
V_{DS}	Drain—Source Voltage	1200	V
V_{GS}	Gate—Source Voltage	−8/+19	
V_{TH}	Gate Threshold Voltage	2.5	
C_{iss}	Input Capacitance	6085	pF
C_{oss}	Output Capacitance	230	
C_{rss}	Reverse Transfer Capacitance	13	
t_r	Rise Time	28	ns
t_f	Fall Time	27	ns
$R_{DS(on)}$	Drain-Source On-State Resistance	16	mΩ
$R_{G(int)}$	Internal Gate Resistance	2.6	Ω
I_D	Continuous Drain Current	115	A
Qrr	Reverse Recovery Charge	604	nC
gm	Transconductance	53	S

4.1. SiC MOSFET Device

In order to facilitate future experiments based on this simulation, this simulation used the existing silicon carbide semiconductor module in the laboratory as the test equipment. The detailed parameters of the tested modules in the experiment are shown in Table 1. Figure 4 shows the device.

Figure 4. C3M0016120D device.

4.2. Simulation Modeling

To improve the simulation, we chose LTspice as the simulation software. As shown in Figure 5, and according to Figure 1, a simulation circuit was built. The main purpose of this simulation was to verify the effect on the switching performance of the OSS and compare it with other switching strategies. It was firstly particularly important to verify whether it was necessary to study the OSS in depth. A simulation can obtain more accurate results and save time. Once the excellent switching performance of the OSS was confirmed, the next step was to build and experiment with actual circuits. According to (24), (26), and data from Table 1, we calculated an optimized delay time t_{d1} of 21 ns and t_{d2} of 41 ns; t_{vd1} and t_{vd2} used the rise time (t_r) and fall time (t_f) in the device data sheet for the simulation, respectively.

Table 2 shows the parameters of some components in the analog circuit. Among them, C_1 and C_4 represent C_{ds}, C_2 and C_5 represent C_{dg}, and C_3 and C_6 represent C_{gs}. In order for the oscillation in the simulation results to be closer to the actual situation, L_S, L_D, and L_G were adjusted.

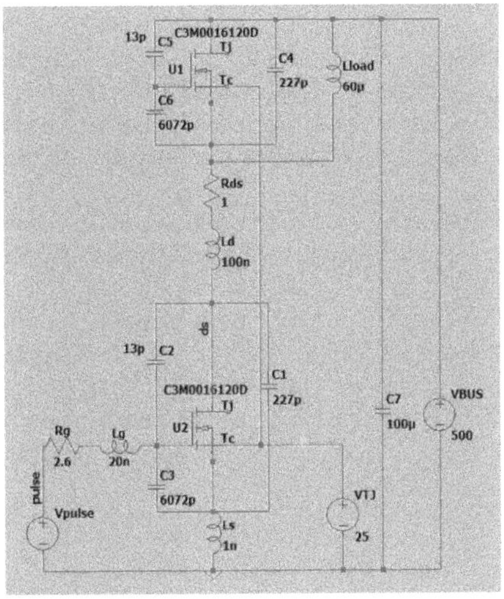

Figure 5. Simulation circuit in Ltspice.

Table 2. Parameters of the SiC MOSFET and associated circuit in LTspice.

Parameters	Value
C_1, C_4	227 pF
C_2, C_5	13 pF
C_3, C_6	6072 pF
C_7	100 µF
L_D	100 nH
L_G	20 nH
L_S	1 nH
L_{load}	60 µH
$R_{DS(on)}$	1 Ω
R_G	2.6 Ω
V_{BUS}	500 V
V_{TC}	25 V

Because the double-pulse simulation of the SiC MOSFET was carried out using the software LTspice, it was necessary to simulate the double-pulse signal, as shown in Figure 6. Because of the convenience of the simulation software, the voltage waveform from the voltage source could be edited directly. When the SiC MOSFET was simulated with the OSS, the waveform could also be adjusted directly at the voltage source, which required corresponding devices in the actual experiments.

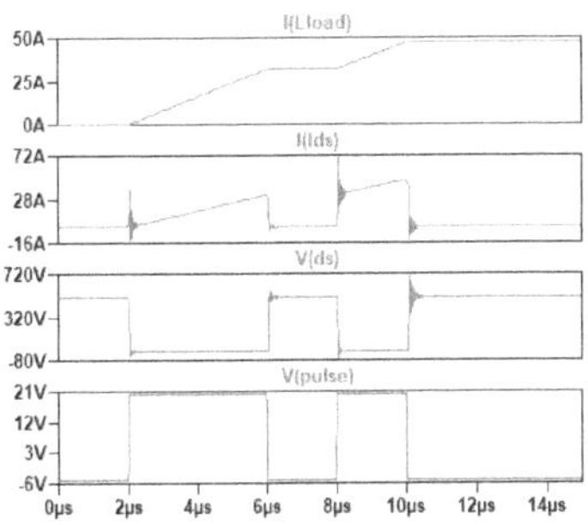

Figure 6. Switching waveforms for the SiC MOSFET with double-pulse switching.

4.3. Simulation Results

In order to verify the feasibility of the proposed OSS, the simulation was run with the gate resistance 2.6 Ω, both at the turn-on stage and turn-off stages; the dc bus voltage was 500 V and the drain current was 30 A. The voltage drop delay times t_{d1} of 21 ns and t_{d2} of 41 ns were calculated by Equations (11) and (12). Meanwhile, as visible from Table 1, the t_{vd1} was 28 ns and the t_{vd2} was 27 ns.

It can be seen in Figure 7a that when the turn-on signal arrived, first, a high V_{GG} (i.e., 20 V) was applied to the gate terminal. After the delay time t_{d1}, which was calculated to be 21 ns, the I_{DS} started to rise and the V_{GG} dropped to 15 V. After the delay time t_{vd1} (i.e., 28 ns), the high V_{GG} restarted to charge the SiC MOSFET. At the turn-off stage, the SiC MOSFET was first discharged by −5 V. After the delay time t_{d2}, which was calculated to be 41 ns, the V_{DS} started to rise from 0 V and the V_{GG} became 0 V. After the delay time t_{vd2} (i.e., 27 ns), the V_{DS} reached the peak value and the negative V_{GG} was discharged again.

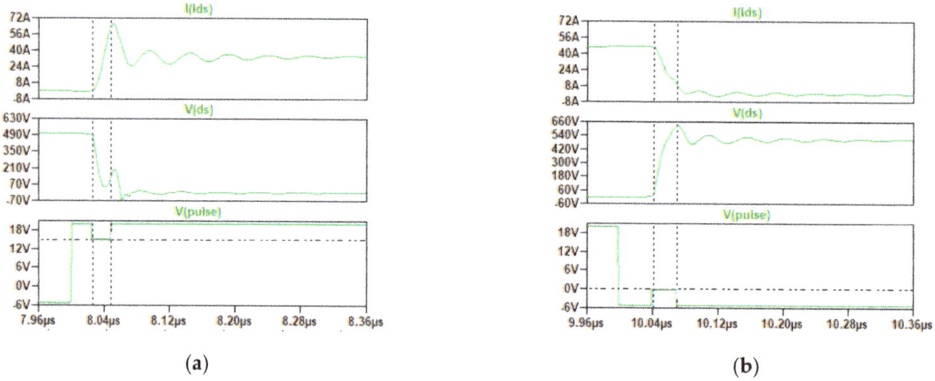

Figure 7. Switching waveforms for the SiC MOSFET with the OSS. (**a**) Turn-on stage, (**b**) turn-off stage.

As mentioned in Section 2, the voltage drop delay times, t_{d1} and t_{d2}, are very important because they can affect the switching performance of the SiC MOSFET. In order to verify the relationship between the delay time and the performance of the switch, and to further the accuracy of the calculation results in the previous chapter, more experiments were carried out. As shown in Figure 8, around the calculation result (i.e., t_{d1} = 21 ns t_{d2} = 41 ns) at an interval of 10 ns, a total of four other data points were taken before and after each transition and were simulated. Then, the simulation results were compared.

For the turn-on transition, because the calculation result of t_{d1} was 21 ns, the five testing data points were 11 ns, 21 ns, 31 ns, 41 ns, and 51 ns, respectively. The reason t_{d1} was not set to 0 was that this waveform was equal to the switching strategy of the digital gate driver. It can be seen from Figure 8a that there was no obvious difference and the V_{DS} and I_{DS} waveforms generated by the five different t_{d1} were not notably different. When t_{d1} was 11 ns, the overshoot of the I_{DS} was the smallest but the overshoot produced by the V_{DS} was the largest.

For the turn-off transition, since the calculation result of t_{d2} was 41 ns, the five test data points were 31 ns, 41 ns, 51 ns, 61 ns, and 71 ns, respectively. It can be seen from Figure 8b that there was an obvious difference and the V_{DS} and I_{DS} waveforms produced by the five different t_{d1} showed little difference. When t_{d1} was 41 ns, the overshoot and oscillation of the I_{DS} were the smallest; the overshoot and oscillation of the V_{DS} were also the smallest.

As the delay time increased or decreased, the calculation results in both cases were the best. However, a trade-off between the V_{DS} and I_{DS} could be achieved when a t_{d1} of about 21 ns at turn on and t_{d2} of about 41 ns at turn off were, respectively, applied.

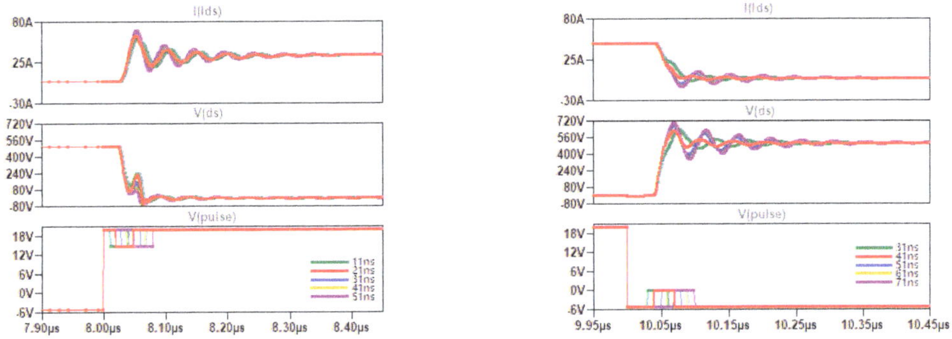

(**a**) (**b**)

Figure 8. The switching performance of the SiC MOSFET with different gate voltage drop time delays. (**a**) Turn-on stage, (**b**) turn-off stage.

4.4. Comparison with Other Switching Strategies

After determining the optimal time delay of the OSS, it was necessary to compare the OSS with other switching strategies. First, we compared the OSS with the hard-switching (HS) strategy to confirm its advantages and improvements. Then, we compared it with a digital gate driver (DGD), which is widely used in teaching and research, to further verify the advantages of the OSS. A detailed experimental comparison of these different switching strategies follows.

4.4.1. Compared to the Hard-Switching Strategy

In order to demonstrate the excellent switching performance of the OSS in terms of overshoot and oscillation, more simulations were performed and the results were compared with the switching performance of hard switching. It can be seen from Figure 9a that during the turn-on stage, the peak value of the I_{DS} dropped from 70.04 A to 60.23 A, a decrease of 14%, and the oscillation was also significantly reduced. However, the overshoot of the V_{DS} increased, which will increase the power loss; although, this will not affect the peak value of the V_{DS}.

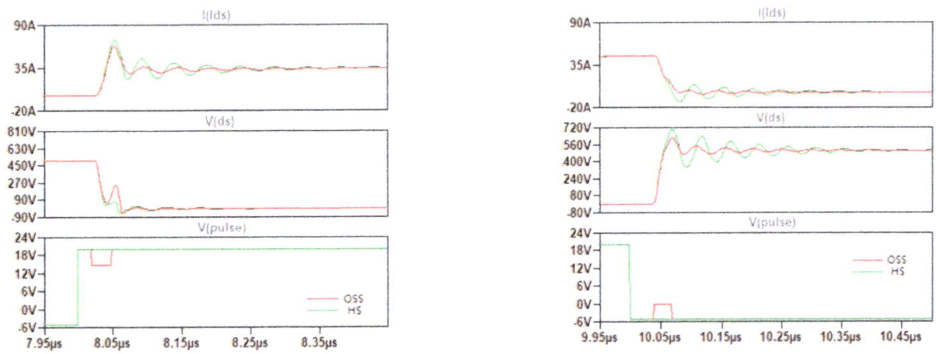

(**a**) (**b**)

Figure 9. Switching performance comparison of the SiC MOSFET between HS and the OSS. (**a**) Turn-on stage, (**b**) turn-off stage.

Unlike in the turn-on process, the OSS successfully reduced the oscillation and peak value during the turn-off process, not only regarding the I_{DS} but also the V_{DS}. It can be seen from Figure 9b that the overshoot of the V_{DS} reduced from 704.56 V to 577.52 V, a drop of 18.03%.

Combining the two processes of turn on and turn off, it can be clearly seen that the OSS effectively reduced the overshoot and oscillation of the I_{DS} and V_{DS} in the SiC MOSFET switch, thereby reducing the overshoot and power consumption of the entire circuit.

4.4.2. Compared to Digital Gate Driver

The digital gate drive system from Agileswitch is a product with a wide range of applications. It can control the voltage drop and voltage drop time of the gate voltage without limitation to affect the switching performance of semiconductors. The control strategy of the DGD is also very advanced. However, the existing DGD control strategy is still unable to design more complex gate drive waveforms, especially the voltage drop after a delay time, like an AGD. Therefore, it was necessary to compare the OSS with existing digital gate driver strategies.

Figure 10 shows the V_{DS} and I_{DS} comparison of the OSS and DGD at the same turn-on time. Compared with the hard-switching strategy, the overshoot of the DGD was reduced by 19%, which was better than the OSS's 14%. Although the DGD was slightly better at reducing the I_{DS} overshoot and oscillation, the overshoot of the V_{DS} was larger, which seriously increased the power losses of the circuit. Although the two switching strategies were both effective in reducing overshoot, they increased power consumption and were low in cost performance. This is why the first generation of DGDs cannot adjust the pulse shape during turn on.

Figure 10 shows the comparison of the V_{DS} and I_{DS} for the OSS and DGD at the same turn-off time. Compared with the hard-switching strategy, the DGD had no advantage and even increased power consumption due to the longer switch-off process.

In order to better compare two different switching strategies, the closing time of the DGD was extended from 68 ns to 88 ns and was then simulated and compared. The new waveform comparison is in Figure 11. Compared with Figure 10b, the DGD effectively suppressed the overshoot and oscillation of the V_{DS} and the overshoot was reduced by 11%. Compared with the OSS, the DGD was inferior. Continuing to extend the turn-off time of the DGD will make the optimization more effective; but, this will inevitably increase power consumption and reduce the frequency of the semiconductor device.

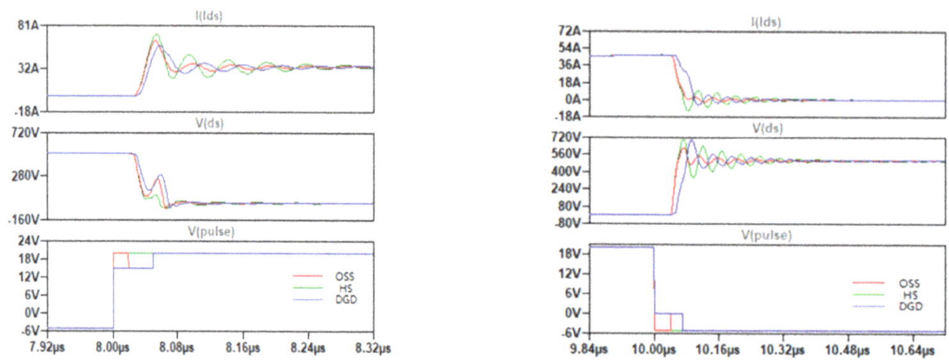

(**a**) (**b**)

Figure 10. Switching performance comparison of the SiC MOSFET between the DGD, HS, and OSS. (**a**) Turn-on stage, (**b**) turn-off stage.

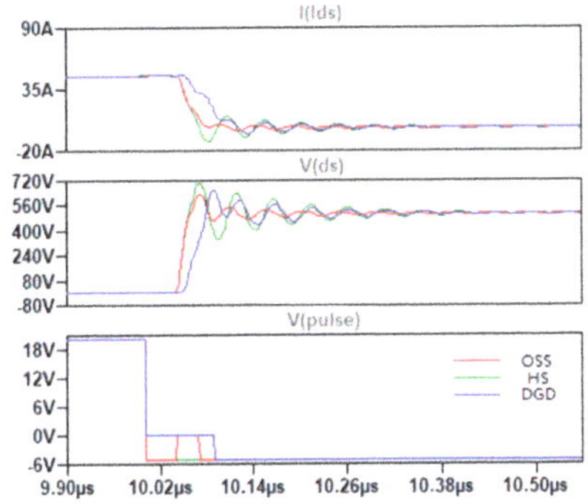

Figure 11. Turn-off switching performance comparison of the SiC MOSFET between the DGD, HS, and OSS when the the DGD time delay was 88 ns.

4.5. Comparison with Different Drop Voltages

The OSS can become more complex and diverse. The previous chapter mainly considered the influence of the voltage time delay on switching performance; this section attempts to study different voltage drops.

It can be seen from Figure 12 that, whether in the on state or the off state, the larger the voltage drop, the less overshoot. Because the increase in the voltage drop reduces the slope of the current and voltage, this proves that the formula in Section 3 is valid. However, during the turn-off process, with the increase in the voltage drop, the oscillation of the voltage and current was obviously increased.

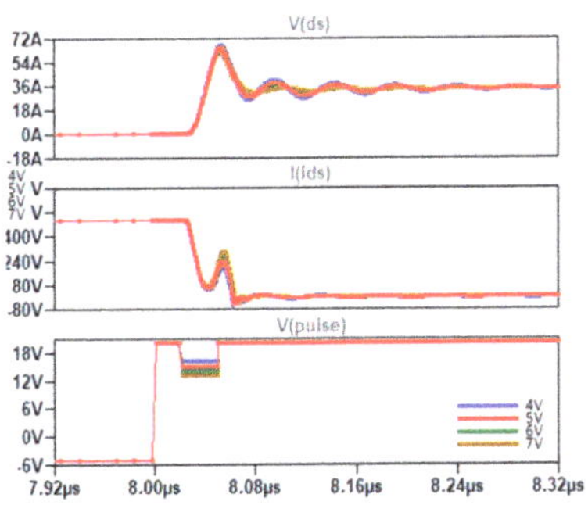

(**a**)

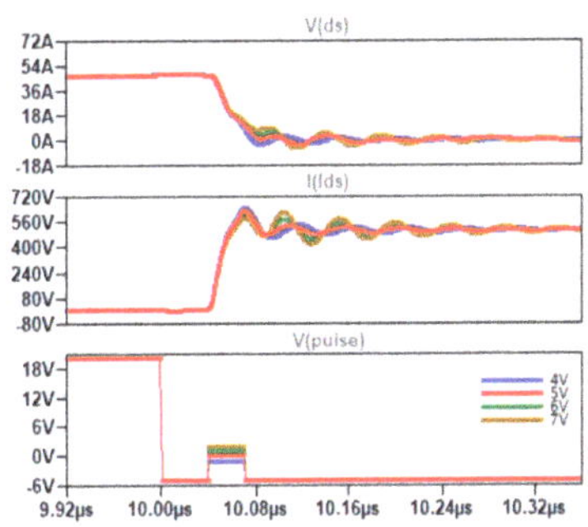

(**b**)

Figure 12. Switching performance of the SiC MOSFET with different drop voltages. (**a**) Turn-on stage, (**b**) turn-off stage.

4.6. Comparison with Different Voltage Drop Times

In the previous simulation, in order to simplify the complexity of the simulation, the voltage drop time (t_{vd}) used the rise time (t_r) and fall time (t_f) in the device datasheet. However, in the OSS, the voltage drop time is also an important parameter, similar to the delay time and voltage drop. Therefore, in this subsection, the effect of the voltage drop time on the switching performance was studied by simulation.

For the turn-on transition, because the t_r was 28 ns, five testing data points were chosen: 18 ns, 28 ns, 38 ns, 48 ns, and 58 ns. It can be seen from Figure 13a that different voltage drop times did not have much influence on the switching performance at the turn-on stage.

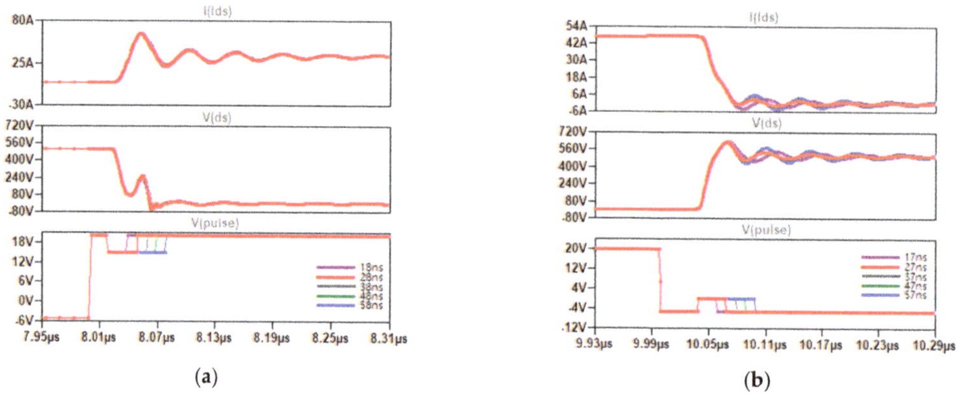

Figure 13. Switching performance of the SiC MOSFET with different voltage drop times. (**a**) Turn-on stage, (**b**) turn-off stage.

For the turn-off transition, because the t_f was 27 ns, five testing data points were chosen: 17 ns, 27 ns, 37 ns, 47 ns, and 57 ns. It can be seen from Figure 13b that in the turn-off stage, different voltage drop times had an impact on the switching performance. When the voltage drop time was 27 ns, the switching achieved the best performance. As the voltage drop time increased, the suppression effect of the OSS on voltage oscillations began to weaken. On the contrary, as the voltage drop time decreased, the voltage oscillation was suppressed; but, a larger reverse current was generated.

4.7. Power Loss Comparison

To determine whether a switching strategy is good or bad, in addition to observing whether it effectively reduces overshoot and oscillation, another important criterion is whether power consumption increases. Due to the advanced nature of LTspice, the product of current and voltage can be quickly integrated directly from the simulated circuit. Table 3 records the power losses from simulation models in one turn-on and turn-off round.

Table 3. Power losses for the simulations in LTspice.

Subsection	Switching Strategy	Power Losses (μJ)
4.4	HS	494.43
	OSS	520.21
	DGD	704.52
	DGD (extended)	681.03
4.5 Different voltage drop	Voltage drop 7 V	675.11
	Voltage drop 6 V	604.05
	Voltage drop 5 V	536.85
	Voltage drop 4 V	473.72
4.6 Different voltage drop time	Voltage drop time −10 ns	491.56
	Voltage drop time 0 ns	539.26
	Voltage drop time +10 ns	582.92
	Voltage drop time +20 ns	606.19
	Voltage drop time +30 ns	686.85
4.3 Different voltage drop delay time	Voltage drop delay time −10 ns	559.25
	Voltage drop delay time 0 ns	518.93
	Voltage drop delay time +10 ns	582.87
	Voltage drop delay time +20 ns	658.26
	Voltage drop delay time +30 ns	743.88

It can be seen from Table 3 that, compared with HS and the DGD, the power consumption of the OSS was slightly higher than that of HS. However, combined with Figure 11, the optimization effect of the OSS remained obvious.

In the comparison of different voltage drops, it can be concluded that the smaller the voltage drop, the smaller the power consumption. However, combined with Figure 12, it can be seen that the smaller the voltage drop, the smaller the suppression effect of the OSS on the circuit overshoot. Similarly, in the comparison of different voltage drop times, it can be concluded that the shorter the voltage drop time, the smaller the power consumption. However, combined with Figure 13, it can be seen that the smaller the voltage drop, the smaller the effect of the OSS on circuit oscillation suppression.

Finally, in the comparison of different voltage drop delay times, it can be concluded that when the voltage drop time is the calculated value, the power consumption is close to the minimum.

5. Conclusions and Discussion

In this paper, an optimal-switching strategy was proposed for improving the switching performance of high-power SiC MOSFETs under hard-switching conditions. In addition, considering the trade-off between switching loss and switching overshoot, the voltage drop delay time was analyzed and calculated. By optimizing the voltage drop delay time of the turn-on and turn-off stages, the OSS can effectively minimize overshoot and suppress oscillation. The simulation results showed that the OSS can reduce the current overshot at the turn-on stage and minimize the voltage overshoot at the turn-off stage. In addition, compared with the hard-switching strategy, the current overshoot in the turn-on phase decreased by 14% and the voltage overshoot in the turn-off phase decreased by 18.03%. This showed that the OSS achieved a more comprehensive control strategy and better switching performance. Compared with digital gate drivers, although each has its own advantages, the OSS is more advanced and suffers less power losses.

The next step in this research is to make the actual circuit and test it. The contribution of this paper is limited due to simulation constraints. The optimization of the OSS in this paper was limited to the control voltage drop delay time; the control strategy could not be fully tested. Only the influence of a single variable of the OSS on the switching performance was tested. In future studies, multivariable control, such as delay time, voltage drop, voltage drop time, etc., could be tested in an actual circuit and jointly controlled to optimize semiconductor switching performance. However, we can confirm that the proposed OSS is promising compared to other methods.

Author Contributions: Methodology, J.T.; validation, J.T.; writing—original draft preparation, J.T.; writing—review and editing, Z.Z. All authors have read and agreed to the published version of the manuscript.

Funding: This research received no external funding.

Data Availability Statement: The data are not publicly available.

Conflicts of Interest: The authors declare no conflict of interest.

References

1. Castellazzi, A.; Fayyaz, A.; Romano, G.; Yang, L.; Riccio, M.; Irace, A. SiC power MOSFETs performance, robustness and technology maturity. *Microelectron. Reliab.* **2016**, *58*, 164–176. [CrossRef]
2. Biela, J.; Schweizer, M.; Waffler, S.; Kolar, J.W. SiC versus Si Evaluation of Potentials for Performance Improvement of Inverter and DC-DC Converter Systems by SiC Power Semiconductors. *IEEE Trans. Ind. Electron.* **2011**, *58*, 2872–2882. [CrossRef]
3. Oswald, N.; Anthony, P.; McNeill, N.; Stark, B.H. An Experimental Investigation of the Tradeoff between Switching Losses and EMI Generation with Hard-Switched All-Si, Si-SiC, and All-SiC Device Combinations. *IEEE Trans. Power Electron.* **2014**, *29*, 2393–2407. [CrossRef]
4. Hazra, S.; Ankan, D.; Cheng, L. High Switching Performance of 1700-V, 50-A SiC Power MOSFET Over Si IGBT/BiMOSFET for Advanced Power Conversion Applications. *IEEE Trans. Power Electron.* **2016**, *31*, 4742–4754.
5. Chen, H.; Divan, D. High speed switching issues of high power rated silicon-carbide devices and the mitigation methods. In Proceedings of the IEEE Energy Conversion Congress and Exposition (ECCE), Montreal, QC, Canada, 20–24 September 2015.
6. Wang, J.; Chung, H.S.; Li, R.T. Characterization and Experimental Assessment of the Effects of Parasitic Elements on the MOSFET Switching Performance. *IEEE Trans. Power Electron.* **2013**, *28*, 573–590. [CrossRef]

7. Sayed, H.; Zurfi, A.; Zhang, J. Investigation of the effects of load parasitic inductance on SiC MOSFETs switching performance. In Proceedings of the IEEE International Conference on Industrial Technology (ICIT), Toronto, ON, Canada, 22–25 March 2017.
8. Nayak, P.; Hatua, K. Modeling of switching behavior of 1200 V SiC MOSFET in presence of layout parasitic inductance. In Proceedings of the IEEE International Conference on Power Electronics, Drives and Energy Systems (PEDES), Trivandrum, India, 14–17 December 2016.
9. Liu, T.; Ning, R.; Wong, T.Y.; Shen, Z.L. Modeling and analysis of SiC MOSFET switching oscillations. *IEEE J. Emerg. Sel. Top. Power Electron.* **2016**, *4*, 747–756. [CrossRef]
10. Yang, B.; Ge, Q.; Zhao, L.; Zhou, Z.; Cui, D. Influence of parasitic elements of busbar on the turn-off voltage oscillation of SiC MOSFET half-bridge module. In Proceedings of the IECON 2017—43rd Annual Conference of the IEEE Industrial Electronics Society, Beijing, China, 29 October–1 November 2017.
11. Fabre, J.; Ladoux, P.; Piton, M. Characterization and implementation of Dual-SiC MOSFET modules for future use in traction converters. *IEEE Trans. Power Electron.* **2015**, *30*, 4079–4090. [CrossRef]
12. Rujas, G.; Lopez, V.M.; Mir, L.; Nieva, T. Gate driver for high power' SiC modules: Design considerations, development and experimental validation. *IET Power Electron.* **2018**, *11*, 977–983. [CrossRef]
13. Liang, M.; Li, Y.; Chen, Q.; Lu, Y. Research on an improved DC-side snubber for suppressing the turn-off overvoltage and oscillation in high speed SiC MOSFET application. In Proceedings of the IEEE Energy Conversion Congress and Exposition (ECCE), Cincinnati, OH, USA, 1–5 October 2017.
14. Jahdi, S.; Alatise, O.; Ortiz Gonzalez, J.A.; Bonyadi, R.; Ran, L.; Mawby, P. Temperature and switching rate dependence of crosstalk in Si-IGBT and SiC power modules. *IEEE Trans. Ind. Electron.* **2016**, *63*, 849–863. [CrossRef]
15. Zhang, Z.; Wang, F.; Tolbert, L.M.; Blalock, B.J. Active Gate Driver for Crosstalk Suppression of SiC Devices in a Phase-Leg Configuration. *IEEE Trans. Power Electron.* **2014**, *29*, 1986–1997. [CrossRef]
16. Riazmontazer, H.; Mazumder, S.K. Optically Switched-DriveBased Unified Independent dv/dt and di/dt Control for Turn-Off Transition of Power MOSFETs. *IEEE Trans. Power Electron.* **2015**, *30*, 2338–2349. [CrossRef]
17. Yamaguchi, K.; Sasaki, Y.; Imakubo, T. Low loss and low noise gate driver for SiC-MOSFET with gate boost circuit. In Proceedings of the IECON 2014—40th Annual Conference of the IEEE Industrial Electronics Society, Dallas, TX, USA, 29 October–1 November 2014.
18. Riazmontazer, H.; Rahnamaee, A.; Mojab, A.; Mehrnami, S.; Mazumder, S.K.; Zefran, M. Closed-loop control of switching transition of SiC MOSFETs. In Proceedings of the IEEE Applied Power Electronics Conference and Exposition (APEC), Charlotte, NC, USA, 15–19 March 2015.
19. Nayak, P.; Hatua, K. Active gate driving technique for a 1200 V SiC MOSFET to minimize detrimental effects of parasitic inductance in the converter layout. *IEEE Trans. Ind. Appl.* **2017**, *54*, 1622–1633. [CrossRef]
20. Paredes, A.; Ghorbani, H.; Sala, V.; Fernandez, E.; Romeral, L. A new active gate driver for improving the switching performance of SiC MOSFET. In Proceedings of the IEEE Applied Power Electronics Conference and Exposition (APEC), Tampa, FL, USA, 26–30 March 2017.
21. Yang, Y.; Wen, Y.; Gao, Y. A Novel Active Gate Driver for Improving Switching Performance of High-Power SiC MOSFET Modules. *IEEE Trans. Power Electron.* **2019**, *34*, 7775–7787. [CrossRef]
22. Katada, R.; Hata, K.; Ymauchi, Y.; Wang, T.W. Digital Gate Driving (DGD) is Double-Edged Sword: How to Avoid Huge Voltage Overshoots Caused by DGD for GaN FETs. In Proceedings of the IEEE Energy Conversion Congress and Exposition (ECCE), Vancouver, BC, Canada, 10–14 October 2021.
23. Miyazaki, K.; Abe, S.; Tsukuda, M.; Omura, I.; Wada, K.; Takamiya, M.; Sakurai, T. General-purpose clocked gate driver IC with programmable 63-level drivability to optimize overshoot and energy loss in switching by a simulated annealing algorithm. *IEEE Trans. Ind. Appl.* **2017**, *53*, 2350–2357. [CrossRef]
24. Anthony, P.; McNeill, N.; Holliday, D. High-Speed Resonant Gate Driver with Controlled Peak Gate Voltage for Silicon Carbide MOSFETs. *IEEE Trans. Ind. Appl.* **2014**, *50*, 573–583. [CrossRef]
25. Chen, L.; Peng, F.Z. Switching loss analysis of closed-loop gate drive. In Proceedings of the Applied Power Electronics Conference and Exposition (APEC), Palm Springs, CA, USA, 21–25 February 2010.
26. Regnat, G.; Jeannin, P.O.; Frey, D.; Ewanchuk, J.; Mollov, S.V.; Ferrieux, J.P. Optimized power modules for silicon carbide MOSFET. *IEEE Trans. Ind. Appl.* **2018**, *54*, 1634–1644. [CrossRef]
27. Baliga, B.J. *Fundamentals of Power Semiconductor Devices*; Springer: Berlin/Heidelberg, Germany, 2008.
28. Liao, X.; Li, H.; Yao, R.; Huang, Z.; Wang, K. Voltage Overshoot Suppression for SiC MOSFET-Based DC Solid-State Circuit Breaker. *IEEE Trans. Compon. Packag. Manuf. Technol.* **2019**, *9*, 649–660. [CrossRef]
29. Lobsiger, Y. Closed-Loop IGBT gate driver and current balancing. Ph.D. Thesis, Department of Information Technology and Electrical Engineering, ETH-Zurich, Seedorf, Switzerland, 2014.

energies

MDPI

Review

Review of Surface Charge Accumulation on Insulators in DC Gas-Insulated Power Transmission Lines: Measurement and Suppression Measures

Fangwei Liang [1], Hanhua Luo [1], Xianhao Fan [1,*], Xuetong Li [1] and Xu Wang [2]

1 Department of Electrical Engineering, Tsinghua University, Beijing 100084, China; liangfangwei@tsinghua.edu.cn (F.L.); lhh22@mails.tsinghua.edu.cn (H.L.); li-xt19@mails.tsinghua.edu.cn (X.L.)
2 School of Mechanical Electronic & Information Engineering, China University of Mining & Technology, Beijing 100084, China; zqt2200404117@student.cumtb.edu.cn
* Correspondence: xianhao_fan@163.com

Abstract: Gas-insulated power transmission lines (GILs) can replace cables and overhead transmission lines, playing an important role in DC transmission systems. However, the influence of surface charge accumulation on insulation reliability cannot be ignored as the operational voltage of the DC GIL increases. In this paper, the measurement methods for the insulator surface potential are summarized, including, dust maps, the Pockels effect method, and the electrostatic probe method. Then, a typical surface charge inversion algorithm is introduced. The main influencing factors of surface charge accumulation are analyzed, such as the applied voltage, insulation gas, insulator shape, and temperature. The charge accumulation pathway is revealed. Furthermore, methods for inhibiting the accumulation of surface charges and promoting the dissipation of accumulated charges are introduced to reduce the surface charges on insulators. Finally, the development direction of DC GIL insulators is predicted. We anticipate that the online monitoring of surface charge distribution, clarifying the percentage of charge accumulation pathways, and optimizing the insulator casting process will be the research directions for the insulator surface charge topic in the future. This article provides a comprehensive understanding of the surface charges of GIL insulators and a reference for the insulation design of DC GILs.

Keywords: gas-insulated power transmission line (GIL); surface potential measurement; surface charge inversion algorithm; surface charge accumulation mechanism; surface charge suppression measure

1. Introduction

Primary energy is mainly distributed in the northwest of China, while most electric energy is consumed in the central and eastern regions. Therefore, it is necessary to implement high-capacity and long-distance power transmission to meet the increasing demand for electricity [1–3]. Compared with AC, high-voltage DC transmission has advantages such as low losses, a large transmission capacity, and convenience for grid interconnection. DC transmission has been widely used in the long-distance transmission of electric energy [4–6]. Ultra-high voltage (UHV) and high-voltage direct current (HVDC) power transmission systems have rapidly developed since the 1950s [7,8]. The DC system is of great significance for China to optimize its energy allocation [9,10]. However, the HVDC transmission lines are facing new challenges, such as high altitudes, large drops, harsh meteorological conditions, low electromagnetic radiation, and strict environmental protection rules [11,12], and the overhead transmission lines and power cables cannot meet these requirements. Under such power supply requirements, gas-insulated transmission lines (GILs) with a SF_6 or SF_6/N_2 gas mixture as the insulation medium have many advantages, such as a large transmission capacity, small footprint, high operational stability, and environmental friendliness [13–15].

GILs can be directly buried underground or installed in tunnels. Therefore, GILs are an excellent choice for large-scale power transmission in addition to overhead transmission lines and power cables [16].

DC GILs will play an important role with the development of DC transmission [17–19]. However, the GILs in operation are basically a AC systems. There are only a few DC GILs due to the difference between AC and DC insulation systems [20,21]. This is because of the different electric field distributions caused by the surface charge accumulation on the insulator [22]. For AC systems, the distribution of the electric field depends on the permittivity. However, the DC system is mainly determined by the resistivity of the material [23,24]. The electric field distribution gradually transitions from the initial capacitive electric field to a steady resistive electric field due to surface charge accumulation [25,26]. Furthermore, the internal insulation system of a DC GIL consists of insulators and compressed gas (e.g., SF_6 or a SF_6/N_2 gas mixture) [16,27]. However, the SF_6 in GILs is one of the greenhouse gases limited by the Kyoto Protocol. Its greenhouse effect is 23,500 times greater than that of CO_2 [28]. The arc extinguishing capability of insulation gas is not required. This is because there are no switchgears in GILs [16]. Furthermore, the SF_6/N_2 gas mixture has the advantages of good insulation strength, stable chemical properties, and lasting mixing characteristics. The SF_6/N_2 gas mixture has been widely used in second-generation GILs to reduce the use of SF_6 [29]. Moreover, the cost can be further reduced due to the low price of N_2. For example, a 20% SF_6/80% N_2 gas mixture has 69% of the insulation strength of pure SF_6 under the same gas pressure. The pressure of the gas mixture needs to be increased by 45% in order to achieve the same insulation strength. However, the use of SF_6 is reduced by 71% [16]. Therefore, the impact of greenhouse gases has been reduced. As a result, the insulator has become the weakest component in DC GILs [30,31].

The insulator is the core component of a GIL [32]. GIL insulators are mainly divided into basin insulators and tri-post insulators. High-voltage conductors are supported by insulators. The conductor is mounted on the axis of the GIL. Additionally, high-voltage conductors are kept insulated from the ground by the insulator [33–36]. The gas seal of the pipeline is achieved by a basin insulator with a sealing ring. The installation position of the tri-post insulator can be easily adjusted. This is because the grounded electrode of the tri-post insulator has a wheel. GIL insulators are mainly composed of epoxy resins with high resistivity [37–39]. As a result, the accumulated surface charge cannot dissipate for a long time. Therefore, the operation of the GIL is threatened due to the flashover triggered by the accumulated surface charges [40–43].

The critical boundary parallel to the insulator surface was selected as the flashover pathway by Kumara et al. The line originated from the maximum electric field on the cathode. A flashover voltage model considering the influence of surface charges was established by determining the initiation, development, and maintenance of the streamer. Under Gaussian distribution, the pulse flashover voltage was decreased by nearly 11% when the maximum positive charge density was 125 $\mu C/m^2$ [44]. Furthermore, the DC $\pm$500 kV gas-insulated switchgear (GIS) developed by the Kansai Electric Power Company, Shikoku Electric Power Company, and Power Development Company of Japan has been successfully applied in the Anan converter station. However, the actual operation voltage is only half of the rated voltage due to the influence of the insulator surface charges [45–47]. This means that the surface charge accumulation of insulators leading to a decrease in the flashover voltage is the issue restricting the development of DC GILs [48–51].

In this paper, the surface charge measurement technology, inversion algorithm, accumulation mechanism, and inhibitory strategies for GIL insulators are summarized. Additionally, the research trends in insulator surface charges are identified. This article provides a reference for the surface charge accumulation of DC GIL insulators.

2. Surface Potential Measurement

The study of surface charges should be based on the accurate measurement of the surface potential distribution on the GIL insulator [52]. The measurement of surface potential is different from that of the voltage for a high-voltage conductor. This is because a certain insulation gap must be maintained between the probe and the measured surface to prevent the leakage of the induced charge onto the probe. At present, the main measurement methods for insulator surface potential include dust maps, the Pockels effect, and electrostatic probes. [53,54].

2.1. Dust Map

Dust maps, also known as Lichtenberg figures, were proposed by Lichtenberg in 1777 [52]. Firstly, the red lead oxide and yellow sulfur are selected. The powder is uniformly suspended in gas. The charged powder with different polarities is absorbed on the insulator surface due to the electrostatic attraction of the surface charges. As a result, the visual analysis of the charge distribution is realized. However, the surface charge distribution cannot be quantitatively measured by a dust map [55]. Since the dust particles are adsorbed on the measured surface, it is not possible to achieve repeated measurements through a dust map [56]. The movement of charged dust particles is also affected by gravity and buoyancy [57]. Furthermore, the original surface charge distribution may be changed. This means that the accuracy of dust maps needs to be further improved. Based on a dust map established by quartz and feldspar powder, the surface charge distribution characteristics with a hollow ring structure are shown in Figure 1 [58].

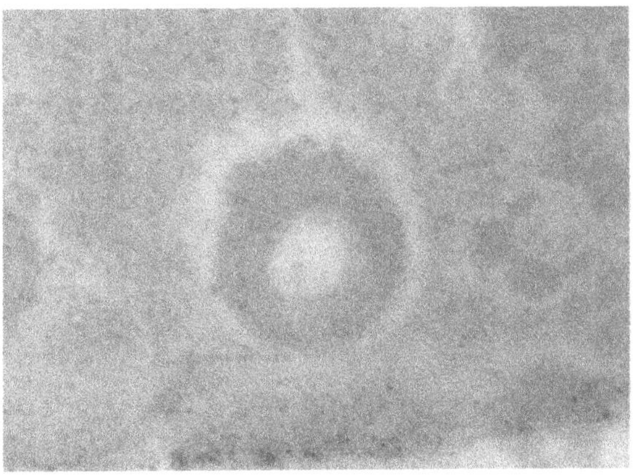

Figure 1. Dust map of surface charge distribution with a hollow ring [58].

2.2. Pockels Effect

The Pockels effect method was proposed by Takada in 1991. The principle was based on the electro-optical effect of Pockels crystals, such as $Bi_{12}SiO_{20}$ and Bi_4GeO_{12} [12,59]. The refractivity of Pockels crystals varies linearly with the electric field [60]. Once the electric field inside the crystals is changed due to the accumulated surface charge, the phase of the incident laser is delayed. The phase delay of light is transformed into a difference in light intensity. Then, the electric field distribution inside the crystal is obtained. As a result, the surface charge density is calculated, as shown in Figure 2 [61]. The real-time and quantitative measurement of surface charges on certain thin and transparent materials can be realized by the Pockels effect [18]. Furthermore, the method is only applicable to the measurement of insulator surface charges under an AC voltage or impulse voltage due to the electrostatic relaxation of Pockels crystals [12].

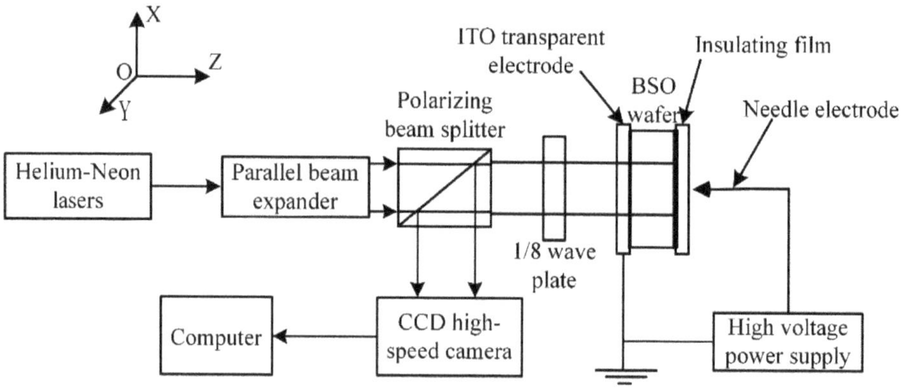

Figure 2. Schematic of the Pockels effect [61].

2.3. Electrostatic Probe

The electrostatic probe method is a non-contact measurement technique [62,63]. There are two commonly used electrostatic probes: capacitive electrostatic probes based on electrostatic induction and Kelvin electrostatic probes designed according to the principle of electric field compensation [64,65]. The capacitive electrostatic probe was proposed by Davies in 1967 [66]. The quantitative measurement of surface charges can be realized by a voltage-following circuit. The data acquisition unit of the probe is composed of a high-input-impedance operational amplifier [67]. In contrast to capacitive probes, Kelvin probes are active probes. Furthermore, Kelvin probes are not affected by the surrounding gas. Kelvin probes are widely used due to their higher accuracy. Their measurement principle is shown in Figure 3 [64].

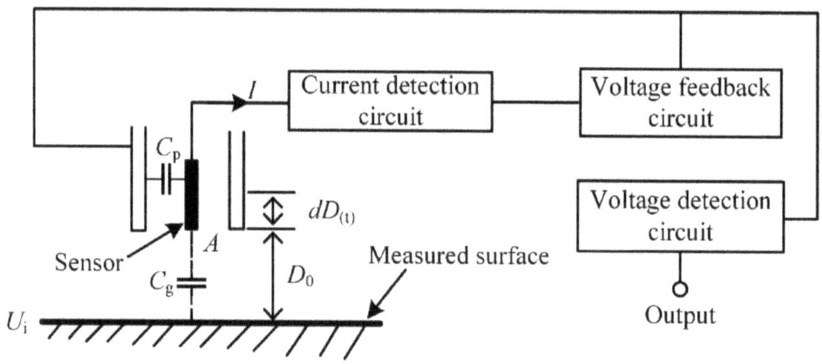

Figure 3. Measurement principle of a Kelvin electrostatic probe [64].

The inductive electrode in the probe is kept vibrating by an oscillator. As a result, the capacitance between the electrostatic probe and the measured surface is changed. If the voltage on the probe is not equal to the measured surface potential, a displacement current is generated on the induction electrode. The voltage on the probe is adjusted until the induced current becomes zero. At this moment, the voltage on the probe is equal to the measured surface potential [68]. Since there is no potential difference between the probe and the measured surface, the partial discharge triggered by the electrostatic probe during the measurement is avoided [57]. This method is an ideal measurement method for surface charges. Typical products include the Trek-341B [69], Trek-347 [70,71], and Monroe 244A electrostatic voltmeter [72,73].

3. Surface Charge Inverse Algorithm

For an electrostatic probe, only the surface potential can be directly measured. Therefore, it is essential to calculate the surface charge density by the inversion algorithm [52]. There have been a number of inversion algorithms proposed in the past, such as the linear algorithm, λ function method, improved λ function method, analytical method, and φ function method [1].

3.1. Linear Algorithm

The linear algorithm was proposed by Davies in 1967. The linear relationship between the surface charge and the induced voltage on the probe is established based on the equivalent measurement circuit, as shown in Figure 4 [1]. The measurement circuit of a capacitance probe consists of two parallel branches, where C_d is the capacitance of the dielectric, C_g is the capacitance formed by the gap between the grounded electrode and the dielectric, C_p is the capacitance between the probe and the measured surface, C_m is the capacitance of the electrostatic probe to the ground, and A is the effective area measured by the probe.

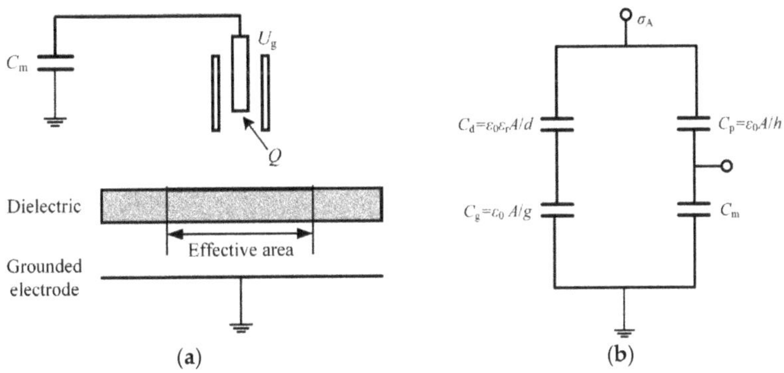

Figure 4. Measurement through capacitive electrostatic probes: (**a**) measurement diagram, (**b**) equivalent circuit [1].

When the gap between the electrostatic probe and the measured surface is small enough, the electric field between them is approximately uniform. Based on the principle of capacitor voltage division, the linear relationship between the surface charge density σ and induced voltage on the probe U_g can be obtained [74].

$$\sigma = M \cdot U_g \tag{1}$$

where M is the scaling coefficient.

The calculation of the linear algorithm is very simple. However, the method is only suitable for thin films smaller than 5 mm. This is mainly because the equivalent circuit is not always consistent during the measurement. Furthermore, the influence of the surface charge outside the measurement area on the results cannot be ignored [67].

3.2. λ Function Method

The relationship between the induced charge on the probe and the surface charges was established by Perdersen et al. based on the λ function [75].

$$q = \iint_{A_0} \lambda \sigma \mathrm{dA} \tag{2}$$

where q is the induced charge on the electrostatic probe, A_0 is the measured area, and λ is the solution of the Laplace equation for the insulation system.

Compared with the linear algorithm, the λ function method has a higher accuracy. However, the method is only suitable for geometrically symmetric insulation systems. Furthermore, the influence of the measurement position of the electrostatic probe is ignored [1].

3.3. Improved λ Function Method

The improved λ function method was proposed by Kumada et al. in 2004 with reference to digital image processing technology [76]. The convolution operation in the spatial domain is transformed by two-dimensional Fourier transform into the product calculation in the frequency domain.

$$\boldsymbol{\Phi}(\mu,\nu) = \boldsymbol{H}(\mu,\nu)\cdot\boldsymbol{\sigma}(\mu,\nu) \tag{3}$$

The calculation of the inversion matrix is avoided. The surface charge density in the frequency domain is solved by division. Finally, the surface charge density σ(x, y) in the space domain is obtained by inverse Fourier transform. Thus, the calculation of the surface charge is greatly simplified. Furthermore, the Wiener filter is used in the frequency domain for noise reduction to further improve the accuracy. However, this method is only applicable to shift-invariant systems, such as large-area flat plates and long cylindrical tubes [77].

3.4. Analytical Method

The analytical method was proposed by Ootera et al. [78]. The method considering all surface charges was based on the theory of electrostatic fields. When the high-voltage electrode is grounded, the insulator surface is divided into n grid units based on the finite element method. Once the mesh is fine enough, the surface charge can be considered uniformly distributed in any cell. According to the superposition theorem of potential, the potential V_i of any element i satisfies the following formula:

$$\begin{bmatrix} P_{11} & P_{12} & \cdots & \cdots & P_{1n} \\ P_{21} & P_{22} & \cdots & \cdots & P_{2n} \\ \vdots & \vdots & & & \vdots \\ \vdots & \vdots & & & \vdots \\ P_{n1} & P_{n2} & \cdots & \cdots & P_{nn} \end{bmatrix} \begin{bmatrix} \sigma'_1 \\ \sigma'_2 \\ \vdots \\ \vdots \\ \sigma'_n \end{bmatrix} = \begin{bmatrix} V_1 \\ V_2 \\ \vdots \\ \vdots \\ V_n \end{bmatrix} \tag{4}$$

$$P_{ij} = \frac{1}{4\pi\varepsilon_0}\int\limits_{S_i} \frac{1}{r_{ij}} dS \tag{5}$$

where ε_0 is the vacuum permittivity; σ_i and S_i are the apparent surface charge density and area, respectively; and r_{ij} is the distance between elements i and j.

Based on the dielectric interface conditions, the relationship between the apparent surface charge density and free surface charge density σ can be written as follows:

$$\begin{bmatrix} F_{11} & F_{12} & \cdots & \cdots & F_{1n} \\ F_{21} & F_{22} & \cdots & \cdots & F_{2n} \\ \vdots & \vdots & & & \vdots \\ \vdots & \vdots & & & \vdots \\ F_{n1} & F_{n2} & \cdots & \cdots & F_{nn} \end{bmatrix} \begin{bmatrix} \sigma'_1 \\ \sigma'_2 \\ \vdots \\ \vdots \\ \sigma'_n \end{bmatrix} = \begin{bmatrix} \sigma_1 \\ \sigma_2 \\ \vdots \\ \vdots \\ \sigma_n \end{bmatrix} \tag{6}$$

$$F_{ij} = \begin{cases} \frac{\varepsilon_2+\varepsilon_1}{2\varepsilon_0} & i = j \\ (\varepsilon_2-\varepsilon_1)\frac{1}{4\pi\varepsilon_0}\int\limits_{S_i} \frac{\boldsymbol{r}_{ij}\cdot\boldsymbol{n}_i}{r_{ij}{}^3} dS & i \neq j \end{cases} \tag{7}$$

where subscripts 1 and 2 represent the insulator side and the gas side, respectively; ε is the permittivity of the material; and $\boldsymbol{n}$ is the unit normal vector in the positive direction pointing from the insulator to the gas.

Therefore, the surface charge density is:

$$\sigma = TV \tag{8}$$

The inversion matrix $\boldsymbol{T}$ is only determined by the material and geometry of the model. Therefore, a surface charge density of 1 $\mu C/m^2$ is defined for the element to solve the surface potential distribution on the insulator. As a result, the inversion matrix $\boldsymbol{T}$ is obtained by moving the unit charge [79].

3.5. φ Function Method

Based on the λ function method, the relationship between the induced voltage on the probe and the surface charge was constructed by Faircloth et al. [80]. The insulator surface is divided into many small regions based on the finite element method. The surface charge of all regions contributes to the surface potential.

$$V = \sigma \boldsymbol{\Phi} \tag{9}$$

For a particular insulation system, the inversion matrix $\boldsymbol{\Phi}$ only depends on the geometric size and material. The $\boldsymbol{\Phi}$ matrix is solved by finite element software, such as ANSYS and COMSOL Multiphysics. Firstly, an inversion calculation model is built. The unit charge is defined on the surface of the measured insulator to calculate the surface potential distribution. As a result, the complete $\boldsymbol{\Phi}$ matrix is solved. The φ function method takes the motion of the electrostatic probe into account, thus improving the accuracy. However, the amount of calculation increases with the dimensions of the inversion matrix.

4. Surface Charge Accumulation Mechanism

The surface charge accumulation of a DC GIL insulator is affected by the applied voltage, insulation gas, insulator shape, temperature, etc.

4.1. Applied Voltage

Under a DC voltage, the surface charge density increases with the applied time. The surface charge distribution is similar under different voltage levels [39]. Furthermore, the polarity of surface charges is reversed due to the increase in the DC voltage when only natural radiation is considered for the generation of charged particles in gas [25]. Since the conductivity of the insulation gas and insulator is a function of the electric field [81], the surface charge accumulation is also affected by the electric field.

4.2. Insulation Gas

In order to reduce the use of SF_6, the insulation gas of second-generation GILs is a SF_6/N_2 gas mixture [16]. Therefore, the surface charge distribution of insulators in 0.5 MPa pure SF_6 and a 20% SF_6/80% N_2 gas mixture were compared by Wang et al. [82]. The results showed that the amount of surface charge was larger in the 20% SF_6/80% N_2 gas mixture under −30 kV. This was mainly because the partial discharge in the gas increased with the decrease in the SF_6 content. Furthermore, the accumulation of surface charges on insulators in the pure SF_6 had a significant polarity effect. That is to say, the surface charges under a negative voltage were obviously larger than those under a positive voltage. However, no significant polarity effect was observed in the 20% SF_6/80% N_2 gas mixture. This was because the pure SF_6 was more sensitive to metal particles. The ionization coefficient of the gas sharply increased due to the enhancement in the local electric fields. As a result, the sources of the surface charges increased [83,84]. For the 20% SF_6/80% N_2 gas mixture, a polarity effect was not observed. This was mainly because the sensitivity of the partial discharge to the electric field distortion was reduced by the N_2.

4.3. Insulator Shape

The accumulation of surface charges is inevitably affected by the geometry of the insulator. Early on, cylindrical insulators were used as simplified models for GIL insulators. Then, cylindrical insulators with shielded electrodes were adopted by Wang et al. [62]. In order to truly reflect the operational conditions of GIL insulators, the circular insulator was proposed by Li et al. [70]. Furthermore, proportionally reduced conical insulators were adopted by Deng et al. [29]. Bowl-shaped insulators were designed to achieve the self-adaptation of surface charges based on nonlinear materials [85]. In addition, an actual GIL basin-type insulator was researched by Qi et al., as shown in Figure 5 [86].

(a) (b) (c) (d)

Figure 5. Different insulator shapes: (**a**) cone-like frustum, (**b**) cone, (**c**) bowl type, (**d**) basin-type [29,70,85,86].

Based on a surface charge transport model considering the microscopic parameters of the gas, the influence of the insulator shape was analyzed by Ma et al. The initial capacitive electric field distribution was compared with the steady-state resistive electric field distribution [87]. The results showed that the surface charge distribution was similar to the normal component of the initial capacitive electric field. The amount of surface charge accumulated on the conical insulator was the largest.

4.4. Temperature

Due to the Joule heating of the operational current, the temperature of the internal conductor in a GIL may rise to 90 °C [88]. Furthermore, the volume conductivity of epoxy insulators is a function of the temperature. The partial discharge in a gas is also affected by the temperature [89]. Therefore, the influence of the temperature on the surface charge accumulation cannot be ignored. The temperature distribution inside a GIL was calculated by Zhou et al. based on thermal convection, radiation, and conduction [90]. The results showed that the temperature distribution inside the GIL gradually decreased from the center conductor to the grounded shell. The temperature gradient was 30 K under a current of 3450 A. The peak of surface charge density was 7.68 nC/cm^2, which increased by 60%. Additionally, high-temperature oil circulation was used to heat the internal conductor of the GIL. As a result, a surface potential measurement platform for DC GIL insulators under a temperature gradient was established, as shown in Figure 6 [91].

Once the temperature gradient was 70 °C, the average value of the surface potential increased from 278 V to 1670 V under a positive voltage. Under a negative voltage, the average surface potential increased nearly twofold. Tang et al. pointed out that the mean free path of electrons increased with the temperature. Therefore, the initial voltage of the corona decreased. As a result, the positive charge density increased under the temperature gradient [41].

The above factors affecting insulator surface charges are mainly due to the differences in the charge accumulation pathway. There are three pathways to accumulating charge on an insulator surface: through the insulator volume, insulation gas, and along the insulator surface [92]. For the ideal case of ignoring the insulation defects inside the GIL, the surface resistivity of the insulator is large enough. The current along the insulator surface is small enough to be ignored. Therefore, the charge accumulated through the insulator volume is uniformly distributed [18]. The conduction current is determined by the volume conductivity of the insulator. For conical insulators, charges of the same polarity accumulate on the insulator surface near the inner conductor [36]. Since space charges are easily generated due to the uneven distribution of volume conductivity, the space charge migrates to the insulator surface, resulting in surface charge accumulation. That is to say, it follows the volume conductivity model proposed by Cooke et al. [12].

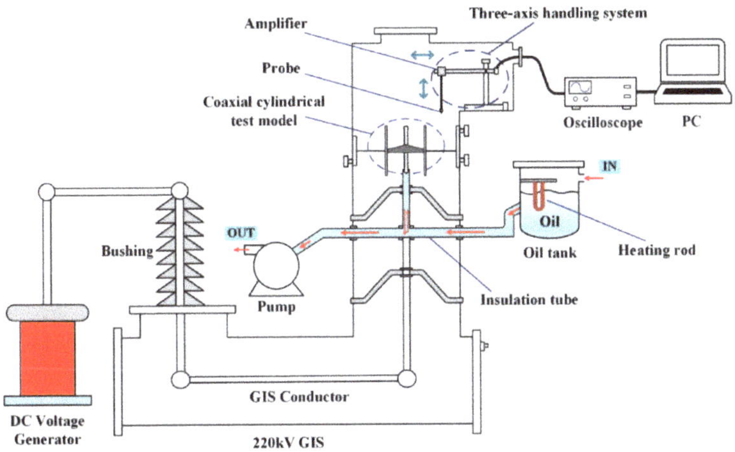

Figure 6. Surface potential measurement system under a temperature gradient [91].

The conduction current through the gas is affected by the generation rate, recombination coefficient, mobility, diffusion coefficient, and other factors [7]. Insulation defects are inevitable within a DC GIL, such as protrusions on the electrode, metal particles on the insulator surface, and the gas-electrode-insulator tri-junction, since the corona discharge of insulation defects is the main cause of surface charge accumulation. The randomly distributed charge spots and charge bands are shown in Figure 7 [18]. This means that the insulator surface charge distribution is uneven [18]. Furthermore, a normal electric field model was proposed by Knecht et al. [93]. It showed that the normal electric field on the insulator surface is the cause of charge accumulation.

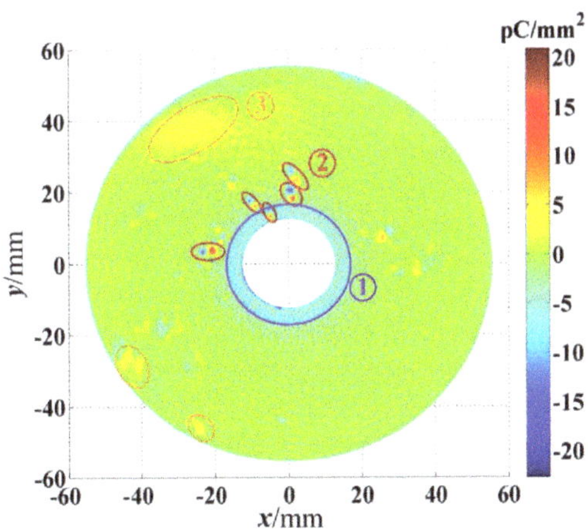

Figure 7. Surface charge distribution pattern: 1. uniform distribution, 2. charge spots, 3. charge clouds [18].

The conduction current along the surface is closely related to the surface conductivity of the material [94]. For insulators with surface modification, the surface charge distribution depends on the conduction current along the surface. The tangential electric field model was proposed by Nakanishi et al. [95]. This model points out that there is an exponential relationship between surface conductivity and tangential electric field, resulting in uneven material surface. Therefore, the discontinuity of conducting current along the surface is the main cause of surface charge accumulation.

The three accumulation pathways of surface charges exist simultaneously. However, the proportion of each part varies significantly under different tests [28]. For the vacuum, corona discharges from the tri-junction of the cathode are the main source of insulator surface charges [96]. Yu et al. pointed out that the surface charge density is closely related to the secondary electron energy emission curve [97]. When the partial discharge in the gas is ignored, the conduction current through the insulator volume is the main accumulation pathway for surface charges [87]. For high-resistivity insulation systems, the accumulation of surface charges is determined by the conduction current in the gas side. The main accumulation pathway gradually changes from the gas to the insulator when the volume resistivity is less than 5×10^{17} Ω ·cm [22]. Furthermore, the generation rate of ion pairs in the gas is one of the key factors. The accumulation pathway changes to the conduction current in the gas with the increase in the ion pair generation rate [48,98].

For basin insulators, a unified surface charge accumulation mechanism was proposed by Li et al. The electric field is divided into three levels based on the leakage current, as shown in Figure 8 [99–101]. For the low electric field, there are many factors, such as residual static charges on the surface, bulk charges inside the insulator, and charged particles in the gas. Each factor may become the core. Thus, charge accumulation under low electric fields can occur in a variety of ways. Under a medium electric field, charges of the same polarity injected through the insulator are the main source. Since the ionization in the gas side is enhanced due to insulation defects under a high electric field, the influence of charged particles in the gas on the surface charge accumulation increases. Therefore, the source of the surface charge is electric-field-dependent.

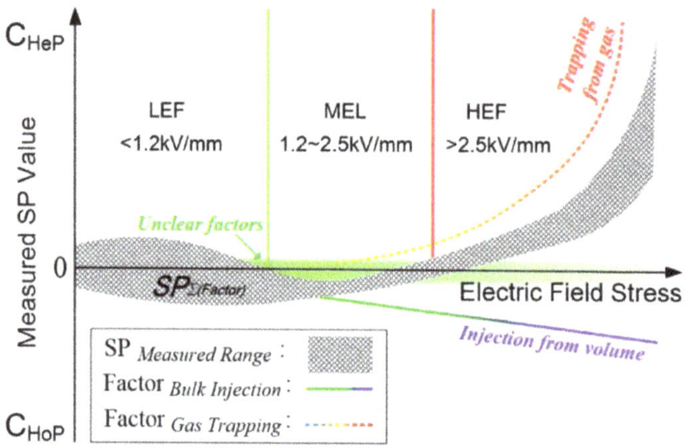

Figure 8. Field-dependent surface charge accumulation model [101].

5. Surface Charge Suppression Strategy

The methods to reduce the surface charge include inhibiting the accumulation of surface charges and promoting the dissipation of accumulated charge, for example, by doping modification, structural optimization, and surface treatment [102]. Since the surface charge accumulation is determined by the conduction current flowing through the insulator when the volume resistivity of the insulator is lower than the critical value of the system [103], increasing the volume resistivity of insulators is an effective method to inhibit surface charge accumulation. The volume conductivity of epoxy resin was reduced through doping with fullerene C60 by Zhang et al. The surface charge on the C60-epoxy composite insulators was reduced due to the hollow cage structure of C60 [104]. That is to say, C60 has extremely strong electrophilic properties. Free electrons in the polymer matrix could be captured.

Nanoparticle doping is the most commonly used modification method for insulation materials, such as Al_2O_3, TiO_2, SiO_2, BN, and AlN [105–112]. This is because a charge accumulation center is introduced by nanoparticles, and the carrier concentration is reduced. As a result, the material conductivity is decreased [113]. Since the surface charge is usually derived from partial discharge in gas for Al_2O_3 epoxy resin insulators [114–116], the normal electric field should be reduced as much as possible during the optimization of insulator shape [87]. For example, a conical insulator was adopted, and the flat bottom was basically parallel to the electric field line. As a result, there was only a small charge on the bottom [36]. Furthermore, a bowl-shaped insulator was designed by He et al. The normal electric field on the insulation area was designed to be negligible in order to reduce the surface charge, as shown in Figure 9 [7]. Therefore, the surface charge accumulation could be reduced by doping modification and structure optimization. However, the influence of the insulation surface is usually ignored. This is because the surface resistivity of GIL insulators is greater than the volume resistivity. The conduction current along the insulator surface is negligible.

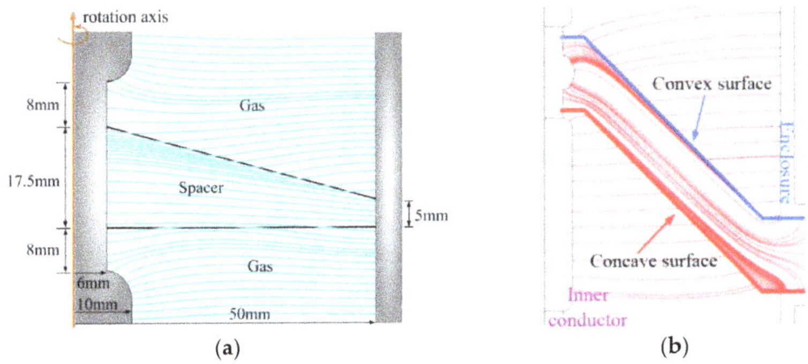

Figure 9. Electric field distribution around insulators: (**a**) cone, (**b**) bowl-type [7,36].

For accumulated surface charges, surface treatment is used to promote dissipation and reduce the surface charge. The main dissipation pathways include: recombination with charged particles in gas, dissipation through the insulator volume, and dissipation along the insulator surface [117]. As partial discharges are random, there is usually little recombination between heteropolar charged particles in gas. For GIL epoxy insulators, the volume conductivity is in the order of 10^{-17} S/m [118]. A surface conductivity of 10^{-18} S has been measured on epoxy resin insulators. As a result, it is impossible for the surface charge to dissipate autonomously for a long period of time. For example, the dissipation curve of the surface potential within 1200 s is approximately horizontal. The dissipation of charge is negligible [119].

The dissipation of surface charges through an insulator material is promoted by the increase in the volume conductivity. However, the insulation size needs to be increased due to the decrease in insulation strength. Therefore, an ideal charge dissipation strategy is to increase the surface conductivity through surface treatment. The surface treatment of insulator materials is used to increase the surface conductivity by 1–2 orders of magnitude, including through surface fluorination, plasma surface treatment, and coating. Thus, the amount of surface charge is reduced by accelerating the dissipation rate.

5.1. Fluoridation

An epoxy resin sheet was fluorinated in a gas mixture of 12.5%F_2/87.5%N_2 by An et al., as shown in Figure 10. A fluorinated layer of micron size was formed. The mobility of the carrier was improved when the fluorination reaction time was less than 60 min. That is to say, the rate of surface charge dissipation along the surface was accelerated [120].

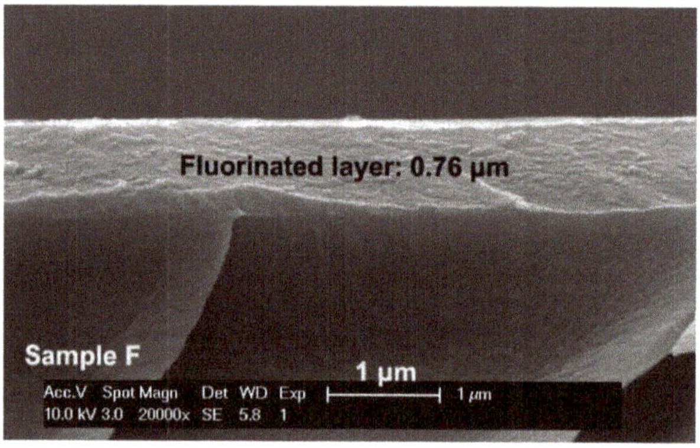

Figure 10. SEM images of the cross-section of fluorinated samples [73].

5.2. Plasma Surface Treatment

Plasma was used by Shao et al. to improve surface conductivity. According to Figure 11, an atmospheric plasma injection device deposited SiOx on the surface of the epoxy resin. Inorganic groups were introduced, such as Si-O-Si and Si-OH [121,122]. Therefore, plasma treatments could also result in shallow trap energy levels on the surface. As a result, the surface charge could rapidly dissipate along the insulator surface [123].

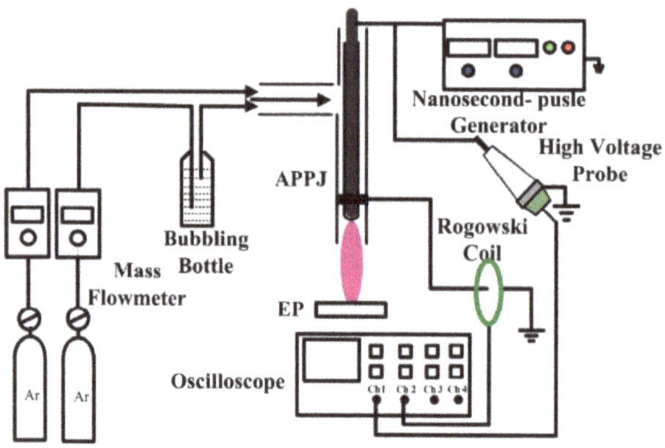

Figure 11. Atmospheric pressure plasma jet device [121].

5.3. Coating

Non-uniform coatings are also functional gradient materials. A functional gradient coating was prepared by Li et al. [124]. The surface conductivity with a gradient was formed by magnetron sputtering ZnO or fluoridation on the epoxy resin surface, as shown in Figure 12.

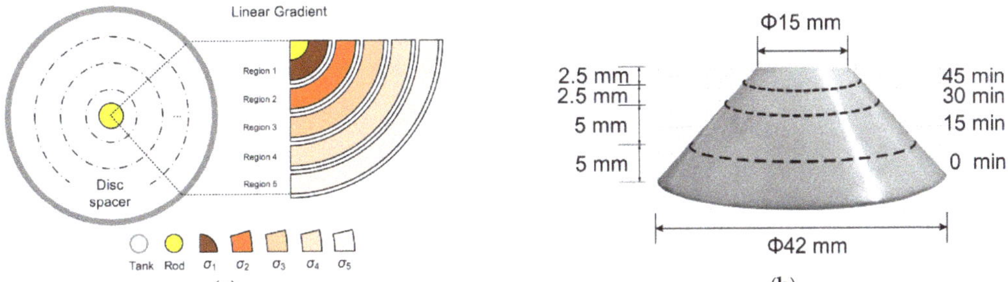

Figure 12. Functional gradient coating on the insulator surface: (**a**) magnetron sputtering, (**b**) fluorination [124,125].

The effect of the concentration and particle diameter of the SiC in the coating on the surface charges was analyzed by Tang et al. [50]. The results showed that the surface trap density was increased due to the SiC/EP coating. Therefore, more surface charges accumulated. Since the surface trap energy level could also be decreased by the coating, the dissipation rate of charges was accelerated. This indicates that the regulation of SiC/EP coatings is attributed to the interaction between the trap density and trap energy level.

6. Conclusions and Suggestions for Future Work

The surface charge measurement methods, charge accumulation mechanisms, and suppression measures for DC GIL insulators were summarized. In order to measure the insulator surface potential, dust patterns, the Pockels effect method, and the electrostatic probe method can be implemented. Then, a surface charge inversion algorithm should be adopted to calculate the surface charge density. Based on the surface charge distribution characteristics, many factors (such as the applied voltage, insulation gas, insulator shape, and temperature) have an impact on the surface charge accumulation. For surface charge suppression, the main solution is inhibiting the accumulation of surface charges and promoting the dissipation of accumulated charges. Finally, suggestions for future research in this area are as follows:

(1) With regard to surface potential measurement, Kelvin electrostatic probes have a high accuracy. However, the application of Kelvin electrostatic probes in online measurements is limited by their lower measurement range. Furthermore, capacitive electrostatic probes have a simple structure and lower cost. Capacitive probes are expected to be applied for the online monitoring of insulator surface charges after the measurement accuracy is further improved.

(2) Focusing on the surface charge inversion algorithm, the environmental noise during the data acquisition cannot be ignored. For online measurement, the changes in the original electric field caused by the probe should be considered to ensure the accuracy of the measurement. Furthermore, a 3D display of the insulator surface charge distribution could be established by the surface charge inversion algorithm combined with data transmission, storage, and image processing.

(3) For the surface charge accumulation mechanism, the proportions of the three pathways of charge accumulation should be clarified. The dominant pathway of charge accumulation should be revealed to provide a theoretical basis for the suppression of surface charges. For the surface charge transport model, the influence of partial discharges in the gas and the surface trap distribution on the insulator cannot be ignored. Additionally, the type of carrier in the gas must be further distinguished rather than simply dividing it into positive ions, negative ions, and free electrons.

(4) In terms of surface charge suppression, the application of coatings on the GIL insulator surface needs to be further tested in terms of adhesion strength and material aging. The effect of coating detachment on insulation should be paid more attention.

Since the material doping modification of an insulator is limited by the casting techniques available, the interface due to the curing properties of different materials should not be ignored. Therefore, understanding the transition region between different doped materials will be one of the challenges for subsequent research.

Author Contributions: Writing—original draft preparation, F.L. and H.L.; writing—review and editing, X.F.; visualization, X.L. and X.W. All authors have read and agreed to the published version of the manuscript.

Funding: This research was funded by the National Natural Science Foundation of China under grant No. 51677061.

Data Availability Statement: The data supporting the findings of this study are available from the corresponding author upon reasonable request.

Acknowledgments: The authors would like to thank Jiayi Liu from Hunan University for the help with the writing.

Conflicts of Interest: The authors declare no conflict of interest.

References

1. Lin, C.J.; Li, C.Y.; Zhang, G.X.; Hu, J.; He, J.L. Review and prospect of surface charge inversion algorithm of Cone-Type spacer based on surface potential measurements. *Proc. CSEE* **2016**, *36*, 6654–6663. (In Chinese)
2. Wang, Q.; Xie, G.; Tian, H.; Peng, Z.; Yang, X. Impact of temperature on the transient DC field distribution of ±1100 Kv UHVDC wall bushing. *IEEE Trans. Power Deliv.* **2021**, *36*, 1449–1457. [CrossRef]
3. Ma, G.; Zhou, H.; Liu, S.; Wang, Y.; Zhao, S.; Lu, S.J.; Li, C.; Tu, Y. Measurement and simulation of charge accumulation on a disc spacer with electro-thermal stress in SF_6 gas. *IEEE Trans. Dielectr. Electr. Insul.* **2018**, *25*, 1221–1229.
4. Wang, J. Research on the Moving Mechanisms of Charged Metal Particles in DC GIL and Suppressing Methods. Ph.D. Dissertation, North China Electric Power University (Beijing), Beijing, China, 2017. (In Chinese).
5. Ma, G.; Zhou, H.; Lu, S.; Wang, Y.; Liu, S.; Li, C.; Tu, Y. Effect of material volume conductivity on surface charges accumulation on spacers under DC electro-thermal coupling stress. *IEEE Trans. Dielectr. Electr. Insul.* **2018**, *25*, 1211–1220. [CrossRef]
6. Liu, Q.; Yan, J.D.; Hao, L.; Zhang, B.; Liu, S. Charge transport and accumulation around a spacer insulator for application in HVDC wall bushing. *IEEE Trans. Dielectr. Electr. Insul.* **2018**, *25*, 281–293. [CrossRef]
7. Li, C.; Lin, C.; Jun, H.; Liu, W.; Li, Q.; Zhang, B.; He, S.; Yang, Y.; Liu, F.; He, J.L. Novel HVDC spacers by adaptively controlling surface charges-Part I: Charge transport and control strategy. *IEEE Trans. Dielectr. Electr. Insul.* **2018**, *25*, 1238–1247.
8. Du, B.X.; Li, J. Effects of ambient temperature on surface charge and flashover of heat-shrinkable polymer under polarity reversal voltage. *IEEE Trans. Dielectr. Electr. Insul.* **2016**, *23*, 1190–1197. [CrossRef]
9. Zhang, Y.L. Observation of Surface Charge for Composite Insulator and Reasearch of Its Influence Factor. Master's Thesis, Hunan University, Changsha, China, 2014. (In Chinese)
10. Zhang, Y.; Lu, L.C.; Yang, X.; Qi, B.; Yang, Z.D.; Dang, Z.P.; Yan, F.Q.; Li, C.R. Exploration on the characteristics of new ceramic materials for insulators used in SF_6 gas insulated HVDC equipment. *Proc. CSEE* **2021**, *41*, 174–182. (In Chinese)
11. Wang, J.; Li, Q.M.; Li, B.T.; Chen, C. Motion analysis of spherical conducting particle in DC GIL considering the influence of inelastic random collisions and SF_6/N_2 gaseous mixture. *Proc. CSEE* **2015**, *35*, 3971–3978. (In Chinese)
12. Zhang, B.Y.; Zhang, G.X. Review of charge accumulation characteristics at gas-solid interface in DC GIL, Part I: Measurement and mechanisms. *Trans. China Electrotech. Soc.* **2018**, *33*, 4649–4662. (In Chinese)
13. Du, B.X.; Du, Q.; Li, J.; Li, A.; Fu, M.L.; Xiao, W. Surface charge dynamic behaviors of Epoxy/BN composite with high thermal conductivity for gas insulated transmission pipeline. *High Volt. Eng.* **2018**, *044*, 2646–2653. (In Chinese)
14. Xue, J.; Li, Y.; Dong, J.; Chen, J.; Li, W.; Deng, J.; Zhang, G. Surface charge transport behavior and flashover mechanism on alumina/epoxy spacers coated by SiC/epoxy composites with varied SiC particle size. *J. Phys. D Appl. Phys.* **2020**, *53*, 155503. [CrossRef]
15. Pan, C.; Tang, J.; Wang, D.; Luo, Y.; Zhuo, R.; Fu, M. Decay characters of charges on an insulator surface after different types of discharge. *Plasma Sci. Technol.* **2017**, *19*, 075503. [CrossRef]
16. Koch, H.; Goll, F.; Magier, T.; Juhre, K. Technical aspects of gas insulated transmission lines and application of new insulating gases. *IEEE Trans. Dielectr. Electr. Insul.* **2018**, *25*, 1448–1453. [CrossRef]
17. Wang, J.; Li, B.T.; Li, Q.M.; Liu, S.H.; Ma, G.M. Impact of linear metal particle on surface charge accumulation of post insulator within DC GIL. *Trans. China Electrotech. Soc.* **2016**, *31*, 213–222. (In Chinese)
18. Xue, J.; Wang, H.; Liu, Y.; Li, K.; Liu, X.; Fan, X.; Deng, J.; Zhang, G.; Guo, B. Surface charge distribution patterns of a truncated cone-type spacer for high-voltage direct current gas-insulated metal-enclosed transmission line/gas-insulated metal-enclosed switchgear. *IET Sci. Meas. Technol.* **2018**, *12*, 436–442. [CrossRef]

19. Gao, Y.; Wang, H.; Yuan, X.; Zhao, H.; Li, Z. Surface charge accumulation on a real size epoxy insulator with bouncing metal particle under DC voltage. *IEEE Trans. Plasma Sci.* **2021**, *49*, 2166–2175. [CrossRef]
20. Gao, Y.; Wang, H.; Li, Z.; Yuan, X.; Zhao, H. Gaseous ionization dependence of surface charge pattern on epoxy insulator with complex surface profile under DC voltage. *IEEE Trans. Plasma Sci.* **2021**, *49*, 1627–1635. [CrossRef]
21. Du, B.X.; Liang, H.C.; Li, J.; Ran, Z.Y. Electrical field distribution along SF_6/N_2 filled DC-GIS/GIL epoxy spacer. *IEEE Trans. Dielectr. Electr. Insul.* **2018**, *25*, 1202–1210. [CrossRef]
22. Lutz, B.; Kindersberger, J. Surface charge accumulation on cylindrical polymeric model insulators in air: Simulation and measurement. *IEEE Trans. Dielectr. Electr. Insul.* **2011**, *18*, 2040–2048. [CrossRef]
23. Zhou, H.; Ma, G.; Wang, Y.; Li, C.; Tu, Y.; Ye, S.; Zhang, B.; Guo, X.; Yang, X. Surface charge accumulation on 500 Kv cone-type GIS spacer under residual DC voltage. *IEEE Trans. Dielectr. Electr. Insul.* **2018**, *25*, 1230–1237. [CrossRef]
24. Winter, A.; Kindersberger, J. Transient field distribution in gas-solid insulation systems under DC voltages. *IEEE Trans. Dielectr. Electr. Insul.* **2014**, *21*, 116–128. [CrossRef]
25. Winter, A.; Kindersberger, J. Stationary resistive field distribution along epoxy resin insulators in air under DC voltage. *IEEE Trans. Dielectr. Electr. Insul.* **2012**, *19*, 1732–1739. [CrossRef]
26. Hoshina, Y.; Yasuoka, T.; Takei, M. Influence of tiny metal particles on charge accumulation phenomena of GIS model spacer in high-pressure SF_6 gas. *IEEE Trans. Dielectr. Electr. Insul.* **2013**, *20*, 1895–1901. [CrossRef]
27. He, S.; Lin, C.; Hu, J.; Li, C.; He, J. Tailoring charge transport in epoxy based composite under temperature gradient using $K_2Ti_6O_{13}$ and asbestine whiskers. *J. Phys. D Appl. Phys.* **2018**, *51*, 215306. [CrossRef]
28. Zhang, B.; Li, X.; Wang, T.; Zhang, G. Surface charging characteristics of GIL model spacers under DC stress in C_4F_7N/CO_2 gas mixture. *IEEE Trans. Dielectr. Electr. Insul.* **2020**, *27*, 597–605. [CrossRef]
29. Wang, H.; Xue, J.Y.; Chen, J.H.; Deng, J.D.; Wan, B.Q.; Zhang, G.J. Influence of metal particles on surface charge accumulation characteristics of GIL spacer in SF_6/N_2 mixture. *Trans. China Electrotech. Soc.* **2018**, *033*, 4663–4671. (In Chinese)
30. Imano, A.M. Accumulation of surface charges on the particle contaminated spacer surface in compressed gas under impulse voltage stress. *J. Electrost.* **2004**, *61*, 1–19. [CrossRef]
31. Li, C.; Xu, Y.; Lin, C.; Chen, G.; Tu, Y.; Zhou, Y.; Lei, Z.; Han, T.; Suraci, S.V.; Wang, J.; et al. Surface charging phenomenon on HVDC spacers in compressed SF_6 insulation and charge tailoring strategies. *CSEE J. Power Energy Syst.* **2019**, *6*, 83–99.
32. Li, C.Y.; Zhang, C.H.; Lv, J.Z.; Liang, F.W.; Liang, Z.D.; Fan, X.H.; Riechert, U.; Li, Z.; Liu, P.; Xue, J.Y.; et al. China's 10-year progress in DC gas-insulated equipment: From basic research to industry perspective. *IEnergy* **2022**, *1*, 400–433. [CrossRef]
33. Du, B.; Dong, J.; Liang, H.; Kong, X. Polarity reversal and over voltage affecting discharge inception of tri-post insulator in ±800 Kv GIL. *IEEE Trans. Dielectr. Electr. Insul.* **2022**, *29*, 223–230. [CrossRef]
34. Li, D.Y.; Zhang, B.Y.; Gao, W.Q.; Hou, Y.C.; Zhang, G.X. "Charge Speckles" pattern in surface charge distribution of insulator under DC voltage. *Trans. China Electrotech. Soc.* **2018**, *33*, 77–83. (In Chinese)
35. Du, Q.D.; Zhang, Q.G.; Zhao, J.P.; Wu, Z.C.; Wang, C.Q. Influence of electric conductivity of materials on field and surface charge distribution of Basin-type insulator. *High Volt. Eng.* **2018**, *44*, 95–101. (In Chinese)
36. Chen, X.; Guan, H.; Jiang, T.; Du, H.; Paramane, A.; Zhou, H. Surface charge dissipation and DC flashover characteristic of DBD plasma treated epoxy resin/AlN nanocomposites. *IEEE Trans. Dielectr. Electr. Insul.* **2020**, *27*, 504–511. [CrossRef]
37. Li, X.; Wan, M.; Zhang, G.; Lin, X. Surface charge characteristics of DC-GIL insulator under multiphysics coupled field: Effects of ambient temperature, load current, and gas pressure. *IEEE Trans. Dielectr. Electr. Insul.* **2022**, *29*, 1530–1539. [CrossRef]
38. Li, J.; Wang, Y.; Chen, H.; Liang, H.; Yao, H.; Zhang, C.; Du, B. Polishing orientation affecting surface charging and flashover characteristics of GIL/GIS epoxy spacer. *IEEE Trans. Dielectr. Electr. Insul.* **2022**, *29*, 1625–1632. [CrossRef]
39. Du, Q.; Zhang, Q.; Wu, Z.; Ma, J.; Zhao, J. Surface charge distribution on DC basin-type insulator. *IEEE Trans. Dielectr. Electr. Insul.* **2019**, *26*, 17–25. [CrossRef]
40. Ma, Y.Y.; Zhang, C.; Kong, F.; Wang, T.T.; Chen, G.Y.; Shao, T. Effect of plasma jet array assisted film deposition on epoxy resin surface electrical characteristics. *High Volt. Eng.* **2018**, *44*, 3089–3096. (In Chinese)
41. Pan, C.; Tiang, J.; Wang, D.; Zhou, R.; Yang, D.; Ye, G.; Fu, M. Influence of temperature on the characteristics of surface charge accumulation on PTFE model insulators. *IEEE Trans. Dielectr. Electr. Insul.* **2017**, *24*, 1210–1219. [CrossRef]
42. Mahmoodi, J.; Mirzaie, M.; Akmal, A.S. Contribution of surface charges on high-voltage DC silicon rubber insulators to DC flashover performance. *IET Gener. Transm. Dis.* **2018**, *12*, 5851–5857. [CrossRef]
43. Amer, M.; Laninga, J.; McDermid, W.; Swatek, D.R.; Kordi, B. Surface charging and its effects on DC flashover strength of insulating materials. *IEEE Trans. Dielectr. Electr. Insul* **2018**, *25*, 2452–2460. [CrossRef]
44. Kumara, S.; Serdyuk, Y.V.; Gubanski, S.M. Simulation of surface charge effect on impulse flashover characteristics of outdoor polymeric insulators. *IEEE Trans. Dielectr. Electr. Insul.* **2010**, *17*, 1754–1763. [CrossRef]
45. Wang, F.; Fang, Z.; Qiu, Y.C. Study of charge accumulation on insulator surface in HVDC gas-insulated switchgear. *Proc. CSEE* **2005**, *25*, 105–109.
46. Tang, J.; Pan, C.; Wang, D.B.; Fu, M.L.; Zhuo, R. Development of studies about surface charge accumulation on insulating material under HVDC. *Trans. China Electrotech. Soc.* **2017**, *08*, 16–27. (In Chinese)
47. Lorenzi, A.D.; Grando, L.; Pesce, A.; Bettini, P.; Specogna, R. Modeling of epoxy resin spacers for the 1 MV DC gas insulated line of ITER neutral beam injector system. *IEEE Trans. Dielectr. Electr. Insul.* **2009**, *16*, 77–87. [CrossRef]

48. Luo, Y.; Tang, J.; Pan, C.; Lin, S.J.; Wang, Z.Q. The transition mechanism of surface charge accumulation dominating way in DC GIS/GIL. *Trans. China Electrotech. Soc.* **2019**, *34*, 185–194. (In Chinese)
49. Chen, K.; Liang, X.D.; Liu, S. Experimental study on the decay process of surface charges on epoxy insulators under different temperature in air. *High Volt. Eng.* **2018**, *044*, 723–1728. (In Chinese)
50. Xue, J.; Chen, J.; Dong, J.; Wang, H.; Li, W.; Deng, J.; Zhang, G. The regulation mechanism of SiC/epoxy coatings on surface charge behavior and flashover performance of epoxy/alumina spacers. *J. Phys. D Appl. Phys.* **2019**, *52*, 405502. [CrossRef]
51. Amer, M.; Laninga, J.; McDermid, W.; Swatek, D.R.; Kordi, B. New experimental study on the DC flashover voltage of polymer insulators: Combined effect of surface charges and air humidity. *High Volt.* **2019**, *4*, 316–323. [CrossRef]
52. Lin, C.; Li, C.; He, J.; Hu, J.; Zhang, B. Surface charge inversion algorithm based on bilateral surface potential measurements of cone-type spacer. *IEEE Trans. Dielectr. Electr. Insul.* **2017**, *24*, 1905–1912. [CrossRef]
53. Xue, J.Y.; Wang, H.; Li, K.F.; Fan, X.F.; Liu, X.Y.; Liu, Y.Q.; Deng, J.B.; Zhang, G.J. Research on charge distribution characteristics on spacer surface in DC GIL. *Proc. CSEE* **2018**, *38*, 6164–6172. (In Chinese)
54. Qi, B.; Gao, C.J.; Xing, Z.L.; Chen, Z.Z.; Zhao, J.K.; Li, C.R. Surface charge distribution on GIS insulator under DC/AC voltage. *IEEE Trans. Dielectr. Electr. Insul.* **2017**, *24*, 3173–3181. (In Chinese) [CrossRef]
55. Li, C.; Zhu, Y.; Zhi, Q.; Sun, J.; Song, S.; Connelly, L.; Li, Z.; Chen, G.; Lei, Z.; Yang, Y.; et al. Dust figures as a way for mapping surface charge distribution-A review. *IEEE Trans. Dielectr. Electr. Insul.* **2021**, *28*, 853–863. [CrossRef]
56. Kumara, S. Surface charges on cylindrical polymeric insulators. *IEEE Trans. Dielectr. Electr. Insul.* **2012**, *19*, 1076–1083. [CrossRef]
57. Zhang, L.; Lin, C.; Li, C.; Suraci, S.V.; Chen, G.; Riechert, U.; Shahsavarian, T.; Hikita, M.; Tu, Y.; Zhang, Z.; et al. Gas-solid interface charge characterization techniques for HVDC GIS/GIL insulators. *High Volt.* **2020**, *5*, 95–109. [CrossRef]
58. Li, C.; Zhu, Y.; Hu, J.; Li, Q.; Zhang, B.; He, J. Charge cluster triggers unpredictable insulation surface flashover in pressurized SF_6. *J. Phys. D Appl. Phys.* **2021**, *54*, 015308. [CrossRef]
59. Wu, K.; Pan, C.; Meng, Y.; Cheng, Y. Dynamic behavior of surface charge distribution during partial discharge sequences. *IEEE Trans. Dielectr. Electr. Insul.* **2013**, *20*, 612–619.
60. Mu, H.B.; Zhang, G.J.; Suzuki, S.; Tanaka, S. Surface charge distribution of polymeric insulating materials under HVAC. *Proc. CSEE* **2010**, *30*, 130–136. (In Chinese)
61. Zhou, W. Study on Influencing Factors and Dissipation Characteristics of Charge Accumulation on the Surface of Insulators under DC Voltage. Master's Thesis, Hunan University, Changsha, China, 2018. (In Chinese)
62. Zhou, H.; Ma, G.; Wang, C.; Wang, G.; Zhang, G.; Tu, Y.; Li, C. Review of charge accumulation on spacer of gas insulated equipment at DC stress. *CSEE J. Power Energy Syst.* **2019**, *6*, 496–517.
63. Takuma, T.; Yashima, M.; Kawamoto, T. Principle of surface charge measurement for thick insulating specimens. *IEEE Trans. Dielectr. Electr. Insul.* **1998**, *5*, 497–504. [CrossRef]
64. Wang, F.; Qiu, Y.; Pfeiffer, W.; Kuffel, E. Insulator surface charge accumulation under impulse voltage. *IEEE Trans. Dielectr. Electr. Insul.* **2004**, *11*, 847–854. [CrossRef]
65. Wang, F.; Liang, F.W.; Chen, S.; Tan, Y.; Zhong, L.; Sun, Q. Surface charge inversion method on cylindrical insulators based on surface potentials measured online. *IEEE Trans. Dielectr. Electr. Insul.* **2021**, *28*, 192–197. [CrossRef]
66. Gao, Y. Study on the Dynamic Characteristics of Surface Charge of Polymer Dielectrics. Ph.D. Dissertation, Tianjing University, Tianjin, China, 2009. (In Chinese)
67. Qi, B.; Gao, C.J.; Xing, Z.L.; Chen, Z.Z.; Zhao, J.K.; Li, C.R. Distribution characteristic for surface charge on GIS insulator under DC/ AC voltage. *Proc. CSEE* **2016**, *36*, 5990–6001.
68. Zhang, Z.; Deng, B.; Li, C.; Li, Q.; Zhang, Z.; Yan, W. Multiphysics coupled modelling in HVDC GILs: Critical re-examination of ion mobility selection. *IEEE Trans. Dielectr. Electr. Insul.* **2019**, *26*, 835–842. [CrossRef]
69. Kindersberger, J.; Lederle, C. Surface charge decay on insulators in air and sulfurhexafluorid-part II: Measurements. *IEEE Trans. Dielectr. Electr. Insul.* **2008**, *15*, 949–957. [CrossRef]
70. Wang, Z.Y.; Wang, J.; Li, Q.M.; Liu, S.H. 3D Simulation and experimental study of metal particles' effect in DC GIL on surface charge accumulation. *Proc. CSEE* **2016**, *36*, 6718–6726. (In Chinese)
71. Du, B.X.; Li, A.; Li, J. Effects of AC and pulse voltage combination on surface charge accumulation and decay of epoxy resin. *IEEE Trans. Dielectr. Electr. Insul.* **2016**, *23*, 2368–2376. [CrossRef]
72. Mallem, H.; Ziari, Z.; Sahli, S. Temperature effect on electrical characteristics of negative DC corona charged polyimide films. *IEEE Trans. Dielectr. Electr. Insul.* **2018**, *25*, 421–427. [CrossRef]
73. An, Z.; Xiao, H.; Liu, F.; Zheng, F.; Lei, Q.; Zhang, Y. Improved resistance of epoxy resin to corona discharge by direct fluorination. *IEEE Trans. Dielectr. Electr. Insul.* **2016**, *23*, 2278–2287. [CrossRef]
74. Wang, F.; Qiu, Y.C.; Zhang, Q.G.; Chen, Q.G. Analysis of factors influence charge accumulation on insulator surface under impulse voltage. *Trans. China Electrotech. Soc.* **2001**, *16*, 51–54. (In Chinese)
75. Rerup, T.O.; Crichton, G.C.; Mcallister, I.W. Using the λ function to evaluate probe measurements of charged dielectric surfaces. *IEEE Trans. Dielectr. Electr. Insul.* **1996**, *3*, 770–777. [CrossRef]
76. Kumada, A.; Okabe, S. Charge distribution measurement on a truncated cone spacer under DC voltage. *IEEE Trans. Dielectr. Electr. Insul.* **2004**, *11*, 929–938. [CrossRef]
77. Kumada, A.; Okabe, S.; Hidaka, K. Resolution and signal processing technique of surface charge density measurement with electrostatic probe. *IEEE Trans. Dielectr. Electr. Insul.* **2004**, *11*, 122–129. [CrossRef]

78. Ootera, H.; Nakanishi, K. Analytical method for evaluating surface charge distribution on a dielectric from capacitive probe measurement-application to a cone-type spacer in ±500 kV DC-GIS. *IEEE Trans. Power Delivery.* **2002**, *3*, 165–172. [CrossRef]
79. Wang, D.B. The Dispersion of Surface Charge of PTFE under DC Voltage and Its Influence on Flashover Characteristics. Ph.D. Dissertation, Chongqing University, Chongqing, China, 2015. (In Chinese)
80. Faircloth, D.C.; Allen, N.L. High resolution measurements of surface charge densities on insulator surfaces. *IEEE Trans. Dielectr. Electr. Insul.* **2003**, *10*, 285–290. [CrossRef]
81. Liang, H.C.; Du, B.X.; Li, J. Electric field regulation and parameter optimization of surface nonlinear conductivity spacer for 500 kV DC-GIL. *IEEE Trans. Dielectr. Electr. Insul.* **2020**, *27*, 1330–1338. [CrossRef]
82. Wang, H.; Xue, J.Y.; Guo, B.H.; Deng, J.; Zhang, G. Comparison of time-dependent variation of surface charge accumulation in SF_6/N_2 and pure SF_6. In Proceedings of the 12th IEEE International Conference on the Properties and Applications of Dielectric Materials, Xi'an, China, 20–24 May 2018.
83. Li, B.T. Study on the Mechanism of Metal Particles on Charge Accumulation on the Surface of Insulators in DC GIL. Master's Thesis, North China Electric Power University (Beijing), Beijing, China, 2016.
84. Guo, C.; Zhang, Q.; You, H.; Ma, J.; Li, Y.; Wen, T.; Qin, Y. Influence of electric field non-uniformity on breakdown characteristics in SF_6/N_2 gas mixtures under lightning impulse. *IEEE Trans. Dielectr. Electr. Insul.* **2017**, *24*, 2248–2258. [CrossRef]
85. Li, C.; Lin, C.; Yang, Y.; Zhang, B.; Liu, W.D.; Li, Q.; Hu, J.; He, S.; Liu, X.; He, J. Novel HVDC spacers by adaptively controlling surface charges-part ii: Experiment. *IEEE Trans. Dielectr. Electr. Insul.* **2018**, *25*, 1248–1258.
86. Qi, B.; Gao, C.J.; Xing, Z.L.; Li, C.R. The accumulation characteristic for surface charge on GIS insulator under switching impulse voltage. *Proc. CSEE* **2017**, *37*, 4538–4547. (In Chinese)
87. Ma, G.; Zhou, H.; Li, C.; Jiang, J.; Chen, X. Designing epoxy insulators in SF_6-filled DC-GIL with simulations of ionic conduction and surface charging. *IEEE Trans. Dielectr. Electr. Insul.* **2015**, *22*, 3312–3320. [CrossRef]
88. Li, C.Y.; Hu, J.; Lin, C.J.; He, J.L. The potentially neglected culprit of DC surface flashover: Electron migration under temperature gradients. *Sci. Rep.* **2017**, *7*, 3271. [CrossRef] [PubMed]
89. Yan, W.; Li, C.; Lei, Z.; Han, T.; Zhang, Z.; Davide, F. Surface charging on HVDC spacers considering time-varying effect of temperature and electric fields. *IEEE Trans. Dielectr. Electr. Insul.* **2019**, *26*, 1316–1324. [CrossRef]
90. Zhou, H.Y.; Ma, G.M.; Liu, S.P.; Li, C.R.; Zhao, S.J.; Tu, Y.P.; Qin, S.C. Study on surface charges accumulation on insulator and its effects on the surface electrical field in DC-GIL with electro-thermal coupling model. *Proc. CSEE* **2017**, *37*, 1251–1259.
91. Zhou, H.Y.; Ma, G.M.; Zhao, S.J.; Liu, S.P.; Li, C.R.; Tu, Y.P.; Song, H.T.; Wu, Z. Effect of temperature on charge accumulation on insulator in DC-GIL. *Proc. CSEE* **2016**, *36*, 6675–6681.
92. Li, C.Y.; Hu, J.; Li, C.J.; Zhang, B.Y.; Zhang, G.X.; He, J.L. Surface charge migration and DC surface flashover of surface-modified epoxy-based insulators. *J. Phys. D Appl. Phys.* **2017**, *50*, 065301. [CrossRef]
93. Li, Z. Simulation of Surface Charge Accumulation on High Voltage DC Composite Insulator. Master's Thesis, Hunan University, Changsha, China, 2014. (In Chinese)
94. Wang, F.; Liang, F.W.; Zhong, L.; Chen, S.; Li, C.; Xie, Y. Short-time X-ray irradiation as a non-contact charge dissipation solution for insulators in HVDC GIS/GIL. *IEEE Trans. Dielectr. Electr. Insul.* **2021**, *28*, 704–709. [CrossRef]
95. Liu, W.J. Design of Surface Charge Test and Measurement System of Insulator at Gas-Solid Interface. Master's Thesis, Hunan University, Changsha, China, 2008. (In Chinese)
96. Suzuki, K.; Kato, K.; Hakamata, Y.; Okubo, H. Real-time and high-speed measurements of charging processes on dielectric surface in vacuum. *IEEE Trans. Dielectr. Electr. Insul.* **2003**, *10*, 563–568. [CrossRef]
97. Yu, K.K.; Zhang, G.J.; Zheng, N.; Raitses, Y.; Fisch, N.J. Monte Carlo simulation of surface-charging phenomena on insulators prior to flashover in vacuum. *IEEE Trans. Plasma Sci.* **2009**, *37*, 698–704.
98. Tian, H.D.; Li, N.Y.; Wu, Z.H.; Guo, Z.; Peng, Z. Surface charge accumulation characteristics of DC GIL column insulators. *Trans. China Electrotech. Soc.* **2018**, *33*, 38–48.
99. Zhang, Z.S.; Wang, Z.M.; Teyssedre, G.; Tohid, S.; Mohamadreza, A.B.; Chen, G.; Lin, C.; Zhang, B.; Uwe, R.; Lei, Z.; et al. Gas-solid interface charge tailoring techniques: What we grasped and where to go. *Nanotechnology* **2021**, *32*, 122001. [CrossRef]
100. Li, Q.; Naderiallaf, H.; Lei, Z.; Wang, Y.; Liu, P.; Zhang, L.; Wang, Z.; Zhang, Z. Surface charge pattern analysis based on the field-dependent charging theory: A review. *IEEE Trans. Dielectr. Electr. Insul.* **2020**, *27*, 257–269. [CrossRef]
101. Li, C.Y.; Lin, C.J.; Chen, G.; Tu, Y.P.; Zhou, Y.; Li, Q.; Zhang, B.; He, J.L. Field-dependent charging phenomenon of HVDC spacers based on dominant charge behaviors. *Appl. Phys. Lett.* **2019**, *114*, 202904. [CrossRef]
102. Zhang, B.Y.; Zhang, G.X. Review of charge accumulation characteristics at gas-solid interface in DC GIL, part II: Charge control and suppression strategy. *Trans. China Electrotech. Soc.* **2018**, *33*, 5–18.
103. Li, C.Y.; Lin, C.J.; Zhang, B.; Li, Q.; Liu, W.D.; Hu, J.; He, J.L. Understanding surface charge accumulation and surface flashover on spacers in compressed gas insulation. *IEEE Trans. Dielectr. Electr. Insul.* **2018**, *25*, 1152–1166. [CrossRef]
104. Zhang, B.Y.; Gao, W.Y.; Hou, Y.; Zhang, G.X. Surface charge accumulation and suppression on fullerene-filled epoxy-resin insulator under DC voltage. *IEEE Trans. Dielectr. Electr. Insul.* **2018**, *25*, 2011–2019. [CrossRef]
105. Wang, S.; Li, A.; Xing, Y.Q.; Li, K.; Li, Z.K. Surface potential and breakdown characteristics of epoxy/AlN nanocomposites under DC and pulse voltages. *IEEE Trans. Appl. Supercond.* **2019**, *29*, 1–5. [CrossRef]
106. Tu, Y.P.; Zhou, F.W.; Cheng, Y.; Jiang, H.; Wang, C.; Bai, F.J.; Lin, J. The control mechanism of micron and nano SiO_2/epoxy composite coating on surface charge in epoxy resin. *IEEE Trans. Dielectr. Electr. Insul.* **2018**, *25*, 1275–1284. [CrossRef]

107. Gao, Y.; Li, Z.Y.; Zhao, N.; Wang, M.H.; Han, T.; Liu, Y. Surface charging on epoxy/Al_2O_3 nanocomposites under DC voltage superimposed by repetitive pulses. *IEEE Trans. Plasma Sci.* **2019**, *47*, 3256–3265. [CrossRef]
108. Gao, Y.; Li, Z.Y.; Wang, M.; Du, B.X. Magnetic field induced variation in surface charge accumulation behavior on epoxy/Al_2O_3 nanocomposites under DC stress. *IEEE Trans. Dielectr. Electr. Insul.* **2019**, *26*, 859–867. [CrossRef]
109. Du, B.X.; Zhang, J.W.; Gao, Y. Effect of nanosecond rise time of pulse voltage on the surface charge of epoxy/TiO_2 nanocomposites. *IEEE Trans. Dielectr. Electr. Insul.* **2013**, *20*, 321–328. [CrossRef]
110. Du, B.X.; Xiao, M. Influence of surface charge on DC flashover characteristics of epoxy/BN nanocomposites. *IEEE Trans. Dielectr. Electr. Insul.* **2014**, *21*, 529–536. [CrossRef]
111. Du, B.X.; Wang, M.Y.; Li, J.; Xing, Y.Q. Temperature dependent surface charge and discharge behavior of epoxy/AlN nanocomposites. *IEEE Trans. Dielectr. Electr. Insul.* **2018**, *25*, 1300–1307. [CrossRef]
112. Chen, Y.; Wu, J. Investigation on relationship between breakdown strength enhancement of composites and dielectric characteristics of nanoparticle. *IEEE Trans. Dielectr. Electr. Insul.* **2016**, *23*, 927–934. [CrossRef]
113. Du, B.X.; Zhang, J.; Gao, Y. Effects of TiO_2 particles on surface charge of epoxy nanocomposites. *IEEE Trans. Dielectr. Electr. Insul.* **2012**, *19*, 755–762. [CrossRef]
114. Ma, J.; Tao, F.; Ma, Y.; Zhao, K.; Li, H.; Wen, T.; Zhang, Q. Quantitative analysis on the influence of surface charges on flashover of insulators in SF_6. *IEEE Trans. Dielectr. Electr. Insul.* **2021**, *28*, 274–281. [CrossRef]
115. Zhang, B.Y.; Wang, Q.; Qi, J.; Zhang, G.X. Measurement method and accumulation characteristics of surface charge distribution on polymeric material under DC voltage. *Proc. CSEE* **2016**, *36*, 6664–6674.
116. LI, C.Y.; Lin, C.J.; Chen, G.; Zhang, Z.S.; Zhang, L.; He, J.L. Review of gas-solid interface charging phenomena of HVDC spacers. *Proc. CSEE* **2020**, *40*, 2016–2025.
117. Xie, Q.; Liang, S.D.; Fu, K.X.; Liu, L.Z.; Huang, H.; Lu, F.C. Distribution of polymer surface charge under DC voltage and its influence on surface flashover characteristics. *IEEE Trans. Dielectr. Electr. Insul.* **2018**, *25*, 2157–2168. [CrossRef]
118. Nakane, R.; Hayakawa, N.; Okubo, H. Time and space transition of DC electric field distribution based on emerging field by accumulated charge in gas-solid composite insulation structure. *IEEE Trans. Dielectr. Electr. Insul.* **2021**, *28*, 440–447. [CrossRef]
119. Zhang, C.; Ma, Y.Y.; Kong, F.; Wang, R.X.; Ren, C.Y.; Tao, S. Surface charge decay of epoxy resin treated by AP-DBD deposition and direct fluorination. *IEEE Trans. Dielectr. Electr. Insul.* **2019**, *26*, 768–775. [CrossRef]
120. An, Z.; Liu, F.; Tang, Y.; Zheng, F.; Zhang, Y. Resistance of surface fluorinated epoxy resin to corona discharge in SF_6 gas. *IEEE Trans. Dielectr. Electr. Insul.* **2016**, *23*, 3659–3667. [CrossRef]
121. Lin, H.F.; Wang, R.X.; Xie, Q.; Zhang, S.; Shao, T. Rapid surface modification by plasma jet to promote surface charge decaying. *Trans. China Electrotech. Soc.* **2017**, *16*, 262–270. (In Chinese)
122. Hu, D.; Ren, C.Y.; Zhang, C.; Qiu, J.T.; Kong, F.; Shao, T.; Yan, P. Effect of deposited film on the surface insulation characteristics of FEP material by atmospheric pressure plasma jet. *Proc. CSEE* **2019**, *39*, 4633–4640. (In Chinese)
123. Kong, F.; Zhang, S.; Lin, H.F.; Ren, C.Y.; Zhang, C.; Shao, T. Effects of nanosecond pulse voltage parameters on characteristics of surface charge for epoxy resin. *IEEE Trans. Dielectr. Electr. Insul.* **2018**, *25*, 2058–2066. [CrossRef]
124. Li, J.; Liang, H.C.; Du, B.X.; Wang, Z.H. Surface functional graded spacer for compact HVDC gaseous insulated system. *IEEE Trans. Dielectr. Electr. Insul.* **2019**, *26*, 664–667. [CrossRef]
125. Du, B.X.; Ran, Z.Y.; Liang, H.C. Novel insulator with interfacial σ-FGM for DC compact gaseous insulated pipeline. *IEEE Trans. Dielectr. Electr. Insul.* **2019**, *26*, 818–825. [CrossRef]

MDPI AG
Grosspeteranlage 5
4052 Basel
Switzerland
Tel.: +41 61 683 77 34

Energies Editorial Office
E-mail: energies@mdpi.com
www.mdpi.com/journal/energies

www.ingramcontent.com/pod-product-compliance
Lightning Source LLC
Chambersburg PA
CBHW061625210926
51850CB00009B/2439
* 9 7 8 3 7 2 5 8 4 7 8 9 1 *